The Cloning Debate

ISSUES

Volume 144

Editors

Lisa Firth and Cobi Smith

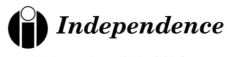

Independence

Educational Publishers
Cambridge

Yeovil College

Y0061325

First published by Independence
PO Box 295
Cambridge CB1 3XP
England

© Independence 2007

Copyright
This book is sold subject to the condition that it shall not,
by way of trade or otherwise, be lent, resold, hired out or otherwise
circulated in any form of binding or cover other than that in which it
is published without the publisher's prior consent.

Photocopy licence
The material in this book is protected by copyright. However, the
purchaser is free to make multiple copies of particular articles for instructional
purposes for immediate use within the purchasing institution.
Making copies of the entire book is not permitted.

British Library Cataloguing in Publication Data
The cloning debate. - (Issues ; 144)
1. Cloning - Moral and ethical aspects
I. Firth, Lisa II. Smith, Cobi
176

ISBN-13: 9781861684103

Printed in Great Britain
MWL Print Group Ltd

Cover
The illustration on the front cover is by
Don Hatcher.

CONTENTS

Chapter One: Human Cloning

Chapter Two: Stem Cell Research

Chapter Three: Animal Cloning

Introduction

The Cloning Debate is the one hundred and forty-fourth volume in the **Issues** series. The aim of this series is to offer up-to-date information about important issues in our world.

The Cloning Debate looks at human cloning, stem cell research and animal cloning.

The information comes from a wide variety of sources and includes:
Government reports and statistics
Newspaper reports and features
Magazine articles and surveys
Website material
Literature from lobby groups
and charitable organisations.

It is hoped that, as you read about the many aspects of the issues explored in this book, you will critically evaluate the information presented. It is important that you decide whether you are being presented with facts or opinions. Does the writer give a biased or an unbiased report? If an opinion is being expressed, do you agree with the writer?

The Cloning Debate offers a useful starting-point for those who need convenient access to information about the many issues involved. However, it is only a starting-point. Following each article is a URL to the relevant organisation's website, which you may wish to visit for further information.

* * * * *

Questions and answers on human cloning

Information from the World Health Organization

Sporadic announcements over the past two years from a number of groups in Asia, Europe, and North America that women under their care have given birth, or are about to give birth, to a 'human clone' have not only produced doubts about the credibility of such claims but have also reactivated the public debate in this area and raised questions about the facts and ethics of what might be involved in human cloning. Public attention has also been stimulated by reports from scientists in several countries that they have either succeeded in creating stem cell lines from 'cloned embryos' or are planning to do so.

The following information is intended to provide background on the topic and the position of the World Health Organization.

What is a clone?

The term clone, from the Greek for 'twig,' denotes a group of identical entities; in recent years, 'clone' has come to mean a member of such a group and, in particular, an organism that is a genetic copy of an existing organism. The term is applied by scientists not only to entire organisms but to molecules (such as DNA) and cells.

Can cloning occur naturally?

Yes, cloning occurs in nature and can occur in organisms that reproduce sexually as well as those that reproduce asexually. In sexual reproduction, clones are created when a fertilized egg splits to produce identical (monozygous) twins with identical DNA. While this is a relatively rare event, many species produce their descendants asexually, that is, without the combining of the male and female genetic material that occurs in sexual reproduction; such offspring are clones of their parent. Since mammals do not normally reproduce asexually, the birth of the lamb Dolly at a research institute in Scotland in 1996 was the first reported mammalian clone produced asexually.

How does 'Dolly type' cloning occur?

To produce Dolly, the researchers used an improved version of the technique of somatic-cell nuclear transfer (SCNT) first used 40 years ago in research with tadpoles and frogs. SCNT begins with an adult somatic cell, for example, a skin cell. 'Adult' means a fully differentiated cell from an organism that had passed the embryonic stage of development, and 'somatic' denotes a body cell (rather than an egg or sperm cell), which possesses the full complement of chromosomes, rather than the half contained in gametes. The nucleus from the somatic cell is transferred to an enucleated egg (that is, one from which the nucleus has been removed). The egg is then activated with electric current or chemicals in order to stimulate it to divide. When the blastocyst stage has been reached, the embryo is transferred into the uterus of a female host, where – if implantation occurs – it can lead to a pregnancy and eventually to the birth of an individual that carries the same nuclear genetic material as the donor of the adult somatic cell. Animals created through SCNT are not precise genetic copies of the donors of their nuclear DNA, however, since a small amount of DNA resides in the mitochondria outside the egg's nucleus; mitochondrial DNA is normally passed on to children only from their mothers. Since a clone would derive its mitochondrial DNA from the egg, not from the donor of the nucleus, the clone and its progenitor would be genetically identical only if the egg came from the progenitor or from the same maternal line.

Would cloning produce identical people?

It is not possible to answer this question with certainty because the experiments required to answer this

question have not been carried out, but experience with mammalian cloning suggests that the answer is 'no'. If successful, SCNT could allow the production of one (or many more) individuals who are, genetically, virtually identical to one another and to the individual whose cell nucleus was used to produce them. This does not mean they would be identical physically or in personality, just as monozygous twins are not identical either, because the development of an organism is influenced by the interaction of its genes and its environment. In the case of human clones, this environment would differ from the moment that each was created, implanted in a uterus, gestated, and born. Furthermore, in all of the species of mammals cloned thus far-including mice, rabbits, pigs and cattle as well as sheep-unpredictable genetic and epigenetic problems have arisen which have not only led to a high rate of abnormalities and prenatal death but have also created health problems for most of the animals born alive, problems which differ from one clone to another.

What justifications are offered for non-reproductive human cloning?

Scientists engaged in cloning for research argue that it presents a unique method for studying genetic changes in cells derived from patients suffering from such diseases as Parkinson's disease, Alzheimer's disease, and diabetes. In February 2004, South Korean scientists reported the creation of a stem cell line from a cloned human embryo. The scientists enucleated 242 oocytes from 16 donors into which they transferred the DNA of ovarian cells from the same donors. Thirty embryos reached the blastocyst stage; from these, the scientists extracted the inner cell mass for the cultivation of stem cell lines, one of which was successfully established. Six months later the U.K. Human Fertilsation and Embryology Authority (HFEA) granted the first license in Europe to allow researchers to use SCNT cloning for embryonic stem cell research. Scientists who are interested in such research look ahead to the day when they believe that embryonic stem cells will be used to assist drug development and evaluation, for diagnostic purposes, and to create cells and tissues for transplantation. For the latter, if the stem cells used in transplantation were derived from embryos cloned from the patient needing the transplant, they might be less subject to rejection than cells, tissues or organs from another person, since the DNA in the cloned cells would be nearly identical to the patient's own. Whether human embryonic stem cells hold unique therapeutic promise-as opposed to stem cells from adult tissues-and, if so, whether the creation of cloned embryos as a source of stem cells would add to their therapeutic value-are matters of ongoing debate in scientific circles.

The term clone, from the Greek for 'twig,' denotes a group of identical entities

---

What uses are suggested for human reproductive cloning?

Proponents of human reproductive cloning argue that it would enlarge the current spectrum of assisted reproductive techniques. In particular, men who do not produce gametes could have children who inherit their genome. In such a case, if the egg came from the wife, the couple would not have to involve a third 'parent' (the sperm donor) in producing their child. Women who do not produce eggs could also have children carrying their genetic information (although they would need a donor egg) and the child would not receive a genetic contribution from the male partner. (In the case of lesbian couples, one might provide the egg, with its mitochondrial DNA, and the other the nuclear DNA). Other reasons offered for using SCNT to create children include: to produce a child with certain genetic features (who could, for example, serve as a bone marrow donor for a diseased sibling); to 'replicate' a deceased child or other loved one; to fulfill the desire for a child based on an admired 'prototype'; or to achieve 'immortality' by living on through one's clone. All of these scenarios raise ethical, legal and social issues.

What ethical arguments have been raised concerning human reproductive cloning?

Although widespread consensus exists internationally among the public, scientists, and policy makers against reproductive cloning, arguments pro and con have been presented. The main arguments brought forward against human reproductive cloning are:

⇨ Physical harm: Experience with animal cloning has shown substantial risks of debilitating and even lethal conditions occurring in the fetuses produced using these techniques; moreover, these problems cannot be individually predicted and avoided at this time. Some of these conditions also present a considerable risk for the gestational mother carrying the cloned animals to term. On the basis of this information, human reproductive cloning would-at this point-constitute a risky experiment that is not sufficiently backed up by successful laboratory and animal research. It would clearly not meet the usual ethical standards in biomedical research. Indeed, the risk-benefit ratio in current preclinical studies of reproductive cloning in animals is so grave that in any other biomedical field, such as the development of a new pharmaceutical product, no responsible researcher would contemplate proceeding to a human trial.

⇨ Research Standards: Responsible biomedical researchers not only engage in thorough laboratory and animal studies before proceeding with human subjects but also submit each step of their work to scientific appraisal through open dissemination of their results in scientific meetings and peer-reviewed journals. Such transparency, which is especially important when scientists operate in private institutions-without day-to-day interaction with colleagues and institutional leaders who are able to bring independent judgment to the design and conduct of the research-has been largely lacking in the human reproductive cloning experiments announced thus far.

⇨ Autonomy: Any child created through SCNT would be unable to give consent to the experiment. Although the same problem arises in any research on the unborn or young children, cloning research is different because, unlike situations in which parents give permission for an experimental intervention that aims to correct an existing problem in a fetus or child, no patient (and hence no medical problem which needs to be remedied) exists prior to the cloning experiment. An issue of autonomy would also arise if a person's DNA were used to create one or more copies without that person's permission or perhaps even without his or her knowledge.

⇨ Conflicts of interest: Special ethical problems would arise when researchers have a financial interest in the outcome of the studies they conduct with human subjects; for this reason, such interests are usually disallowed by ethical standards or, where they are unavoidable, special expectations of openness and independent review of research are required.

⇨ Psychological/social harm: The cloned individual may suffer psychological harm from its status as a 'genetic copy' of somebody else. The clone might be dominated by the person who creates him or her, unduly constrained by expectations based on the abilities or life course of the donor, or stigmatized by society. It remains uncertain whether these concerns can be effectively addressed by education and legislation.

⇨ Dignity: The Universal Declaration on Human Genome and

Human Rights (UNESCO 1997) as well as many other documents state that reproductive cloning is contrary to human dignity. This position is mainly based on the following ethical considerations:

a. cloning is an asexual mode of reproduction, which is unnatural for the human species; a cloned individual will not have two genetic parents; generation lines and family relationships would be distorted.

b. cloning limits the lottery of heredity, which is an essential component in ensuring that each human life (or lives, in the case of monozygous twins) begins as something that has never existed before.

c. cloning furthers an instrumental attitude toward human beings, that is, that people exist to serve purposes set by other people. When cloning is used in this fashion, dignity is undermined in two different but related ways-first, a clone's right to an individual life-course will be constrained by others' expectations that he or she will behave in certain ways (based on experience with the genetic progenitor's life), and second, the clone may not (or may not wish to) behave in those ways, because behavior is not shaped by genes alone, and will hence disappoint others' expectations and suffer the consequences.

d. especially in conjunction with other means of genetic modification, cloning risks turning human beings into manufactured objects; this is not only contrary to human dignity but unwise, as human beings lack the prescience to meddle successfully with evolution and genetic diversity in this fashion.

⇨ Justice: Health care resources should be devoted instead to other health or research needs that address more urgent problems than any associated with reproductive cloning; furthermore, not only are there few if any people for whom reproductive cloning would offer the only means of establishing a family, but if cloning became established as an assisted reproductive technology,

it would probably only be available to a small group of privileged individuals with the financial resources to afford it.

The main arguments brought forward for human reproductive cloning are:

⇨ Beneficence: A new treatment option could be offered to infertile couples. Predetermining the genetic make-up of children would allow selection of desirable traits and bestow advantages on them.

⇨ Autonomy: People should be free in their reproductive decisions; the state or international organizations do not have the right to interfere with reproductive autonomy.

What regulations exist on cloning?

There are a variety of national laws on cloning; many other bills have been submitted and are currently under consideration. As of now, approximately 35 nations have adopted laws forbidding reproductive cloning. Some, including Germany, Switzerland, and some jurisdictions in the United States prohibit all forms of human cloning, whereas others, among them the United Kingdom, China, and Israel and other jurisdictions in the United States prohibit only 'reproductive' cloning, but allow the creation of cloned human embryos for research. International documents, such as the Universal Declaration on Human Genome and Human Rights (UNESCO, 1997) and the World Medical Association's Resolution on Cloning (1997), have addressed the issue, but lack binding legal force. The United Nations discussed an international convention against the reproductive cloning of human beings during the General Assembly in November 2002; the debate on whether to adopt a treaty banning human cloning is still ongoing, with the principal issue being whether the ban should include research as well as reproductive cloning.

⇨ The above information is an extract from information provided by the World Health Organization and is reprinted with permission. Visit www.who.int for more information or to view the full text.

© World Health Organization

What is cloning?

Information from the Roslin Institute

Much confusion happens when people see the word 'clone' used. Depending on the age of the dictionary, the definition of biological cloning can be:

⇨ A group of genetically identical individuals descended from the same parent by asexual reproduction. Many plants show this by producing suckers, tubers or bulbs to colonise the area around the parent.

⇨ A group of genetically identical cells produced by mitotic division from an original cell. This is where the cell creates anew set of chromosomes and splits into two daughter cells. This is how replacement cells are produced in your body when the old ones wear out.

⇨ A group of DNA molecules produced from an original length of DNA sequences produced by a bacterium or a virus using molecular biology techniques. This is what is often called molecular cloning or DNA cloning

⇨ The production of genetically identical animals by 'embryo splitting'. This can occur naturally at the two cell stage to give identical twins. In cattle, when individual cells from 4- and 8-cell embryos and implanted in different foster mothers, they can develop normally into calves and this technique has been used routinely within cattle breeding schemes for over 10 years.

⇨ The creation of one or more genetically identical animals by transferring the nucleus of a body cell into an egg from which the nucleus has been removed. This is also known as Nuclear Transfer (NT) or cell nuclear replacement (CNR) and is how Dolly was produced.

⇨ The above information is reprinted with kind permission from the Roslin Institute. Visit their website at www.roslin.ac.uk for more.

© *Roslin Institute*

Cloning: new horizons in medicine

A clone is a group of cells or organisms which are genetically identical and have all been produced from the same original cell. Identical twins are natural clones, but over the last 50 years we have developed the ability to produce clones artificially. There are three main types of cloning, each with the potential to deliver great medical breakthroughs – but there are some ethical dilemmas attached.

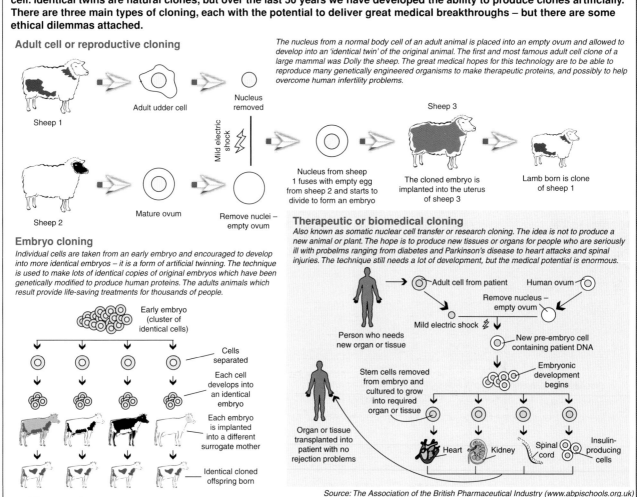

Adult cell or reproductive cloning

The nucleus from a normal body cell of an adult animal is placed into an empty ovum and allowed to develop into an 'identical twin' of the original animal. The first and most famous adult cell clone of a large mammal was Dolly the sheep. The great medical hopes for this technology are to be able to reproduce many genetically engineered organisms to make therapeutic proteins, and possibly to help overcome human infertility problems.

Sheep 1 — Adult udder cell — Nucleus removed

Mild electric shock

Sheep 2 — Mature ovum — Remove nuclei – empty ovum

Nucleus from sheep 1 fuses with empty egg from sheep 2 and starts to divide to form an embryo

Sheep 3 — The cloned embryo is implanted into the uterus of sheep 3 — Lamb born is clone of sheep 1

Embryo cloning

Individual cells are taken from an early embryo and encouraged to develop into more identical embryos – it is a form of artificial twinning. The technique is used to make lots of identical copies of original embryos which have been genetically modified to produce human proteins. The adults animals which result provide life-saving treatments for thousands of people.

Early embryo (cluster of identical cells)

Cells separated

Each cell develops into an identical embryo

Each embryo is implanted into a different surrogate mother

Identical cloned offspring born

Therapeutic or biomedical cloning

Also known as somatic nuclear cell transfer or research cloning. The idea is not to produce a new animal or plant. The hope is to produce new tissues or organs for people who are seriously ill with probelms ranging from diabetes and Parkinson's disease to heart attacks and spinal injuries. The technique still needs a lot of development, but the medical potential is enormous.

Adult cell from patient — Human ovum

Remove nucleus – empty ovum

Person who needs new organ or tissue

Mild electric shock

New pre-embryo cell containing patient DNA

Embryonic development begins

Stem cells removed from embryo and cultured to grow into required organ or tissue

Organ or tissue transplanted into patient with no rejection problems

Heart — Kidney — Spinal cord — Insulin-producing cells

Source: The Association of the British Pharmaceutical Industry (www.abpischools.org.uk)

Human cloning – the ethical issues

Information from the Society, Religion and Technology Project

Dolly the cloned sheep caused a media sensation. But after the hype subsided, what are the real issues? Why would it be wrong to clone human beings? What about possible medical uses of the technology, like cloning embryos for replacement body cells?

What's the Church doing here?

Since 1993, the Church of Scotland's Society, Religion and Technology Project (SRT) has looked in depth at the ethics of genetic engineering and cloning in animals and plants with an expert working group. Leading scientists, including Professor Ian Wilmut, leader of the Roslin team that produced Dolly, discussed issues with specialists in ethics, theology, sociology and risk, which culminated in a major book 'Engineering Genesis', published by Earthscan in 1998. So when Dolly hit the headlines, in February 1997 the church was already in a position to offer a balanced and informed view on this local Edinburgh issue with global implications. In May 1997 the Church of Scotland General Assembly was one of the first organisations in the world to give a formal view on human and animal cloning, which has been much quoted, for example in a UNESCO declaration. SRT has been deeply engaged in UK, European and international ethical discussions first about cloning, and then also on stem cell issues after the isolation of human embryonic stem cells – which added to the complexity. To help shed light on these confused and often misrepresented issues, we have produced three information sheets: on human cloning, animal cloning and embryonic and adult stem cells.

What is cloning?

The word 'clone' comes from a Greek word for taking a cutting from a plant. To clone is simply to make an exact genetic copy of an existing organism. It happens naturally in many plants (if you bury a potato it sprouts clones of itself), and even a few animals. Significantly, it does not normally happen in mammals and humans, except for 'identical' twins. And as we shall see, this is very different from cloning when it comes to the ethical aspects. Dolly changed all that. She is a sheep created by taking cells from the udder of a ewe and 'reprogramming' them to create a new embryo by a process known as nuclear transfer, and implanting the embryo in another ewe. This was a biological revolution. It had been thought impossible to grow a mammal from body tissue. And if it was possible in sheep (and now cattle and mice also), could it be done in humans? And if it could be, should it be?

Cloning runs counter to the evolutionary need to maintain a basic level of genetic diversity

---

What are the ethical objections to cloning human beings?

The overwhelming reaction from most people was that it should not be done, but a fear that someone might try. Statements opposing cloning human beings have issued from numerous national and international organisations, like the UN, the Council of Europe, the European Parliament, the European Commission's ethical advisors, the UK Human Fertilisation and Embryology Authority, many professional medical bodies, and also the scientists at Roslin who cloned Dolly. The UK and many other Governments have now banned it in law. But what exactly is wrong with human cloning? It is not enough that it is unnatural; much medical treatment is also unnatural. The key question is should we respect a biological distinction or celebrate our God-given capacity to override

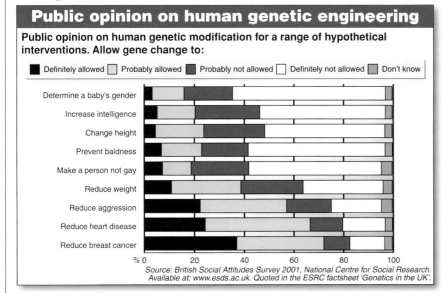

Public opinion on human genetic engineering

Public opinion on human genetic modification for a range of hypothetical interventions. Allow gene change to:

■ Definitely allowed　▨ Probably allowed　▨ Probably not allowed　□ Definitely not allowed　▨ Don't know

- Determine a baby's gender
- Increase intelligence
- Change height
- Prevent baldness
- Make a person not gay
- Reduce weight
- Reduce aggression
- Reduce heart disease
- Reduce breast cancer

% 0　20　40　60　80　100

Source: British Social Attitudes Survey 2001, National Centre for Social Research. Available at: www.esds.ac.uk. Quoted in the ESRC factsheet 'Genetics in the UK'.

it? Four basic reasons have emerged: control, instrumental use of other humans, risk and relationships.

Control and instrumentality

In one sense, cloning runs counter to the evolutionary need to maintain a basic level of genetic diversity, and the variety God has created in nature. What about identical twins then? We don't regard them as any less human. Indeed, we often comment on how different each is, because we are far more than just our genes. But the mere existence of 'identical' twins does not justify the practice of cloning. Ethically, twinning and cloning are as different as chalk and cheese. In twinning an embryo whose genetic composition has never existed, and is unknown at that point, spontaneously splits into two. It is something random, uncontrolled. Cloning would select the genetic composition of some existing person and try and make another individual with the same genes. It is an intentional, controlled act to produce a specific known end. The crucial point is not the genetic identity but the human act of control of it. The most fundamental ethical case against human cloning is that no human being should have their complete genetic make up pre-determined by another human.

Cloning is also instrumental. Parents influence and select for their children in a thousand ways socially, but so far they have never been able to choose their genes for them. We can reject their upbringing, but we cannot change our genes. In most of the cases people have speculated about, like providing a donor for bone marrow for a sibling with leukaemia, the person cloned would not be created for their own benefit but someone else's. This is an instrumental way of using another human, as a means to someone else's ends. again this is unacceptable human control. An exception might be cloning one member of a couple to solve infertility, but this would raise other profound problems.

Risk and relationships

No one knows the psychological effects of discovering one was the twin of one of one's 'parents' or sibling. Am I just a copy of someone else who's already existed and not really 'me'? What would be my relationship to them? Since we have no sure way of knowing in advance, we surely do not have the right knowingly to inflict that risk on another person. Lastly there is the physical risk, in the light of the animal cloning experience. Major pregnancy difficulties are often feature of cloning work in sheep and cattle. Cloned mice seem to die younger. The understanding of the basic science of nuclear transfer is still rudimentary. No one knows how to guarantee that the cell reprogramming process would not lead to serious abnormalities in the offspring or danger to the mother. To translate such risks into humans would be utterly unethical medically. Each new cloned species has produced unexpected results. One cannot 'put down' a deformed cloned baby the way one might a suffering lamb. It is virtually impossible to conceive of a point where one would risk trying cloning on humans.

Parents influence and select for their children in a thousand ways socially, but so far they have never been able to choose their genes for them

---

What about cloned human embryos for research?

But if cloning people is wrong, what about medical applications? In 1997, the Church of Scotland General Assembly called for an international ban on reproductive human cloning, and we support the proposal for such a UN Treaty, but not for a ban on research uses. In principle, cloning research could throw light on cell and embryo behaviour, fertility and ageing. Vital therapies might result, such as in the case of mitochondrial disease. But any applications which would mean producing cloned embryos raise serious ethical questions. The proposal to extract stem cells from cloned embryos to produce genetically matched replacement cells for degenerative diseases remains highly speculative. No one knows if it could be done, the prospects of therapeutic success, the risks, or where huge numbers of human eggs would come from. This so-called therapeutic cloning seems unlikely to be used except in research. For example, one suggestion is to use cloned embryos to create stable lines of disease state cells. For some all embryo research is unethical, and even for those not opposed to embryo research creating cloned embryos poses ethical problems. We explore these issues further in our information sheet 'Embryonic and Adult Stem Cells: Ethical Dilemmas'. Reproductive human cloning should be something no scientist would do, but with mavericks keen to abuse the technology for personal fame, unless and until there is a UN ban, no one should seek to create cloned embryos even for research.

⇨ The above information is reprinted with kind permission from the Church of Scotland's Society, Religion and Technology Project. Visit www.srtp.org.uk for more information.

© Church of Scotland

A humanist discussion of embryo research

Information from the British Humanist Association

Humanist ethics

Humanists seek to live good lives without religious or superstitious beliefs. They use reason, experience and respect for others when thinking about moral issues, not obedience to dogmatic rules. Humanists promote happiness and fulfilment in this life because they believe it's the only one we have. When deciding whether something is right or wrong, humanists consider the evidence and the probable effects of choices.

> ## Medical science has advanced to the point where we have options that were unthinkable even a few years ago and where old rules cannot cope with new facts

Embryo research is a subject that demonstrates the difficulties of rigid unchanging rules in moral decision making. Medical science has advanced to the point where we have options that were unthinkable even a few years ago and where old rules cannot cope with new facts.

21st century medicine could be transformed by research into using human embryos as a source of tissue-repairing cells, often called 'therapeutic cloning'. This is different from 'reproductive cloning', where cloned embryos would be grown from a cell taken from one individual and then implanted in a womb where they would develop into near replicas of their one parent.

Therapeutic cloning

Using 'stem cells' from very early embryos (under 14 days of age), which are capable of developing into

any of the specialised cells of the body, replacement tissue could be grown in the laboratory and used to cure many currently incurable conditions, avoiding the problem of immune rejection. But the use of human embryos raises ethical questions and provokes much opposition, particularly from religious and anti-abortion groups, who use similar arguments to those used to oppose contraception and abortion to object to the exploitation of a living human embryo. Opponents also fear that embryo research for therapeutic purposes is a 'slippery slope' that will lead to the cloning of human beings. Some religious groups accuse medical researchers of 'playing God'.

Humanists respect life, but are not religious and so do not worry about 'playing God' or believe in 'the sanctity of life'. (Human beings have been 'playing God' for a long time, intervening beneficially in reproductive and medical processes.) For humanists the most important consideration in ethical questions on life and death is the quality of life of the individual person. In the case of embryo research, humanists would focus on two issues: whether an embryo is indeed a person, and whether the research on and subsequent use of embryo cells would do more good than harm.

Is an embryo a person?

At the early stage where research is focused, an embryo has few of the characteristics we associate with a person. It is a fertilised human egg, with the capacity to develop into a person, but its cells have not yet begun to form into specialist cells that would form particular parts of the body (which is why they are potentially so useful). There is no brain, no self-awareness (or consciousness), no way of feeling pain or emotion, so an early stage embryo cannot suffer.

Should we consider embryo donors?

Fertility treatments produce many of the 'spare embryos' that would be used, and parents might feel some attachment to these or concern or guilt about what happens to them. It would seem right to inform them fully about what might happen to their embryos, and to take their feelings into account. If they do not consent to donating their embryos for medical research, they should not be used. On the other hand, spare embryos are routinely disposed of at the moment, so already they are not treated as human beings, and parents do not seem unduly concerned. Donors may even prefer their embryos to be used to help someone, rather than wasted, just as many people consent to organ donation.

So, should we allow therapeutic cloning?

If an embryo's cells can be used to alleviate human suffering, the good consequences seem to outweigh the harmful ones, as long as the legal cut off point for research is sufficiently early. Do embryos being produced specially for research and therapeutic purposes by IVF in the laboratory raise any new moral issues? The consequences seem to be much the same, so a humanist would probably think not. So most humanists would support therapeutic cloning, because they do not consider very early embryos to be people, unlike some religious people.

Questions to think about

⇨ Do you think an early stage embryo is a human being with human rights?

⇨ How big is a 14-day old embryo? What does it look like? (Ask your Biology teacher)

⇨ Find out what use(s) might be made of cloned embryo cells.

⇨ Find out what the law on embryo research is now, and if there are any plans to change it. Do you support change? Give reasons for your viewpoint.

⇨ Find out who supports embryo research and who opposes it, and why.

⇨ Do we need more people?

⇨ Do we need more of specific kinds of people?

⇨ Do we need them so much that we are prepared to take risks in order to produce them?

⇨ Do we need more of specific kinds of people? How are you deciding your answers to these questions? What principles and arguments influence your answers?

⇨ How is the humanist view on this issue similar to that of other worldviews you have come across? How is it different?

Is this a slippery slope? Will we be cloning human beings next?

It is often argued that therapeutic cloning will inevitably lead to reproductive cloning – a classic 'slippery slope' argument. Many scientists (over half a small sample polled by The Independent in August 2000) think that therapeutic cloning will develop research techniques and skills that will inevitably be used for human reproductive cloning, There does indeed appear to be a technical slippery slope between therapeutic cloning and reproductive cloning, which has already proved possible in animals (in the case of Dolly the sheep). However, there are also very clear differences between the two, which make it possible to distinguish between them morally – it is not a moral slippery slope. New cures for disease are needed and the consequences of producing new treatments seem, on balance, to be good. But in an over-populated world new ways of creating human beings are not needed, and the consequences of producing human beings by cloning might not be good at all.

Cloned human beings might suffer from unintended physical side-effects of the process, such as premature ageing or infertility, or other abnormalities, and we might decide that it would be unethical to grow human beings experimentally to the point where this could be detected. It is possible that cloned children could suffer psychologically – as do some adopted children or those born as a result of IVF. In many ways cloning seems like a vanity project: a parent would have to be very confident of his own qualities to want to produce a near identical child, and the expectations the parent would have of the child might hinder its healthy emotional development. For these reasons, humanists would probably oppose reproductive cloning.

Some of the fears raised by cloning do seem exaggerated – it would be a costly, slow and unlikely way to raise an super-fit army or Olympic athletes or scientific geniuses – selection from the existing population, training, and education would be much more cost effective. Fears that children would be absolutely identical to their parent, a younger twin, in effect, seem unfounded when we take into account the vital role of environment and upbringing in making us who we are: the child's experience of life would be very different from its parent's.

But even if some fears are irrational, we would need very good reasons to embark on an experiment that we might not fully know the results of for several decades. Knowing how to do something does not mean that we necessarily have to do it. Human beings have all kinds of knowledge and capabilities that we have decided it would be better not to use, for example, the USA and the former USSR have stockpiles of nuclear weapons that could wipe out life on Earth, but have chosen not to deploy them. Reproductive cloning might be another example. It is worth remembering that it not for scientists alone to decide how to use their research – it is a decision for society, and that means all of us.

With thanks to Professor Lewis Wolpert, CBE, FRS

⇨ The above information is reprinted with kind permission from the British Humanist Association. Visit www.humanism.org.uk for more information.

© British Humanist Association

Public opinion on embryo research

Respondents were asked 'British scientists are legally permitted to carry out a limited range of experiments using early human embryos up to 14 days after conception (at which point they are a cluster of about 2,000 cells). Do you believe that it is, or is not, acceptable to use "spare" early embryos left over from fertility treatment, such as IVF, for the purposes of medical research?'

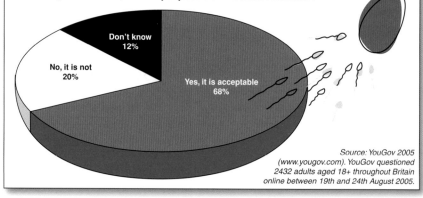

Don't know 12%

No, it is not 20%

Yes, it is acceptable 68%

Source: YouGov 2005 (www.yougov.com). YouGov questioned 2432 adults aged 18+ throughout Britain online between 19th and 24th August 2005.

Biologists want to drop the word 'cloning'

Information from the New Scientist

Don't say cloning, say somatic cell nuclear transfer. That at least is the view of biologists who want the term to be used instead of 'therapeutic cloning' to describe the technique that produces cloned embryos from which stem cells can then be isolated. This, they argue, will help to distinguish it from attempts to clone a human being.

But will it? Kathy Hudson and her colleagues at the Genetics and Public Policy Center in Washington DC asked more than 2000 Americans whether they approved of deriving stem cells from embryos produced by cloning. For half of the sample they used the term 'SCNT' instead of 'cloning', and this raised approval ratings from 29 per cent to 46 per cent, Hudson told a meeting of the American Society of Human Genetics in New Orleans last week.

'Using the term 'SCNT' instead of 'cloning' dramatically raised approval ratings'

SCNT would also be used in any attempt to clone a human being, so Hudson also asked about creating babies using SCNT. This too raised the approval rate from 10 per cent to 24 per cent – which is not what scientists had in mind.
21 October 2006

⇨ Information from the New Scientist. Visit www.newscientist.com for more information.
© *Reed Business Information Ltd*

Animal-human embryos

Questions answered

By Roger Highfield

Why do scientists want to create hybrid animal-human embryos?
Human eggs are in short supply – being also in demand for fertility treatment – and are expensive, costing about £3,000 per woman. Moreover, the procedure carries risks. Rabbit or cow eggs would enable scientists to do basic stem cell research more cheaply and safely.

How will it be done?
The animal egg will be emptied of its nuclear DNA and fused with DNA from a human cell, as was done with Dolly the sheep, or fused with an entire human cell. The latter method will mean that human DNA will also replace the second type of DNA – that in power plants of cells, called mitochondria – to some extent so that the resulting cloned embryos are even more human.

Has this been done before?
Yes. This method has been pioneered by Dr Hui Zhen Sheng's team at the Shanghai Second Medical University, China, where she fused human cells with rabbit eggs to produce early stage embryos, which in turn yielded human stem cells.

What is the result?
These embryos have only one type of nuclear DNA and are called hybrids or 'cybrids.'

Is the Chinese work convincing?
Yes. Dr Sheng's team used donor cells from the foreskins of a five-year-old boy and two men, and facial tissue from a woman. They fused the nuclei (containing DNA) of the human cells with rabbit eggs from which they had removed the nuclei. The resulting embryos were dismantled and used to derive six lines of stem cells, flexible cells that can develop into any type.

Is this regulated?
The use of cybrids is 'a grey area' in terms of current regulations, which are under review. The work raises a basic question – if you take a human nucleus and put it in a rabbit egg, is it a human embryo? Whatever, it is still counted as a human embryo by the fertility watchdog, the HFEA.

So why not use hybrid cells to repair patients?
They would not be used for stem cell treatments of patients because of concerns about disease and because cybrid cells may still contain rabbit DNA in mitochondria, the 'batteries' of the cell. The few rabbit genes present may generate proteins that would be attacked by the human immune system, for example.

Will the cells of the hybrids have other benefits?
Yes. Although cybrid stem cells could not be used for treatments and will not be allowed to develop beyond an early stage, they will prove invaluable for studies of how to clone embryonic stem cells more efficiently – so human eggs could eventually be used – and could be used to test drugs on cell lines created from people with Alzheimer's disease, and shed light on the basic disease process.
8 November 2006
© *Telegraph Group Limited, London 2007*

Chimeras, hybrids and 'cybrids'

Information from the Christian Medical Foundation

By Calum MacKellar

In biotechnology it is now possible to combine elements between organisms of different species. It is also possible to create cloned animals using parts of eggs from one species and nuclear genetic material from another. It is even possible to create novel organisms via interspecies combinations of gametes. Should such procedures ever be permissible between animal species? If so, should we ever combine human beings with animals?

Early in 2007 a parliamentary committee met to discuss the technological and ethical issues surrounding any possible mixing together of human and animal species. They sat as a response to proposed legislation which, if enacted, would have banned such work. After effective lobbying from scientists and other associated interest groups, the committee decided in favour of the creation and limited use of human-nonhuman hybrids, chimeras and 'cybrids'.

The interest in mixing species is neither new, nor is it confined to the realms of myth or fiction. True, many ancient cultures told stories and built statues of entities such as human-lion sphinxes and winged horses, but the natural mixing of animals has occurred for centuries. A mule, for example, starts life when a male donkey mates with a female horse (a cross between a female donkey and male horse is less common, and called a hinney). The gametes (sperm and egg) fuse and the resulting embryo develops into a healthy animal. Though normally infertile, there are even occasional reports of mules giving birth.

But in recent years, research has raised a host of new possibilities. In 1984 scientists created the world's first sheep-goat chimera by fusing a sheep embryo and a goat embryo. The resulting 'geep' consisted of goat cells and sheep cells. Externally this combination was obvious, as the skin that grew from the sheep embryo was woolly, while the areas of skin that originated from goat cells bore hair.

Should we ever combine human beings with animals?

The potential power of inter-species combinations became clearer with a series of experiments conducted in the late 1990s. In these, small sections of brains from quail embryos were transplanted into the developing brains of chickens. When they hatched, the resulting chickens exhibited quail-like vocal trills and head bobs, showing that the transplanted parts of the brain were not only incorporated into the brain, but that such mixing of tissues could allow complex behaviours to be transferred between species.

The next step for many scientists is to start combining human and nonhuman cells. The immediate objective is not to generate beings that are fully grown half-humans, but to create a source of stem cells that could potentially be used in research and therapy. Initial requests for permission to perform this work envisage that any embryo created by mixing human and nonhuman cells would not be allowed to develop beyond the 14-day stage.

Political landscape

This push to develop the combining of human and nonhuman cells has come from technological developments since the 1990 legislation. Anxiety about this new possibility can however be seen in a 2001 UK report from the Home Office's Animal Procedures Committee, which recommended that 'No licences should be issued for the production of embryo aggregation chimeras, especially not cross-species chimeras between humans and other animals, nor of hybrids which involve a significant degree of hybridisation between animals of very dissimilar kinds'.

This reluctance to involve human cells and embryos is also found in European documents such as Article 13 of the Council of Europe's European Convention on Human Rights and Biomedicine, which prohibits any action that aims to modify the human genome in a way that will be passed on to future generations. In effect this would ban any genetic technologies applied at a very early embryonic stage of life. European policy makers are clearly anxious about the technology, though the UK government has not so far signed up to this convention.

In his January 2006 State of the Union address, American President George W Bush expressed his position when he slipped in a small but significant comment that announced his intention to ban human-animal hybrids: 'A hopeful society has institutions of science and medicine that do not cut ethical corners, and that recognize the matchless value of every life. Tonight I ask you to pass legislation to prohibit the most egregious abuses of medical research: human cloning in all its forms, creating or implanting embryos for experiments, creating human-animal hybrids, and buying, selling, or patenting human embryos. Human life is a gift from our Creator – and that gift should never be discarded, devalued or put up for sale.'

This File will examine what is currently possible, and what is envisioned for the near future. By drawing on Christian principles it will ask whether any form of species mixing is ethically justified, particularly where one of the components is human.

What's been done so far

One complication is that there are various ways of deliberately mixing two species of animal (see box). Each process produces a different outcome and raises different issues.

Genes from humans in bacteria
At the simplest end, there are the many examples of genes harvested from the human genome and placed inside bacteria. These transgenic bacteria have huge medical and commercial potential. For example, most insulin is now produced from E. coli with the insulin gene from a human inserted amongst other genes. These bacteria consequently produce an individual human protein, but are far from bearing any distinctively human characteristics.

Genes from humans in mice
Moving up in scale, there are also many thousands of strains of mice that have had sizable pieces of genetic code that originated from the human genome spliced into their genes. Many of these are used in cancer and pharmaceutical research as experimental animals that mimic human disease. In terms of each specific disease they have distinctly humanised traits, but they are still clearly mice.

Andi – primate with jellyfish gene
In the first two examples a small element of human DNA has been incorporated with a mass of another organism's genes. In the case of Andi, the process was the other way around. Andi was the first primate to have a package of foreign genes inserted into its genome. The genes came from jellyfish and although present in his cells, they did not function particularly well. Andi, however, shows the possibility of introducing new genes into primate cells, and thus that it would potentially be possible to add new genes to human beings. If the gene were merely repairing the function of an organ such as the liver, then most people would probably accept this as a legitimate medical intervention. But what would happen if the gene were expressed in the brain and altered the individual's ability to think, or their innate behaviour?

Cow egg-human clone
In 1999 the US company Advanced Cell Technology Inc announced that it had developed a method for producing primitive human embryonic stem cells by uniting human adult material with a cow egg. This egg had previously had its nucleus removed. The company hopes this method will enable them to produce 'unlimited' supplies of stem cells for research into transplant medicine.

Researchers hope the technique will remove a very important barrier in current research into embryonic stem cell transplantation therapies, namely the need for fresh human eggs of which there is a very limited supply for the creation of cloned embryos. Scientists are eager to obtain these embryos in order to harvest their stem cells for biomedical research.

Adult human material has been fused with rabbit eggs to creat rabbit-human hybrid embryos

Rabbit-human hybrid embryos
In August 2003, Hui Zhen Sheng of Shanghai Second Medical University, China, announced that rabbit-human 'cybrid' embryos had been created. Researchers fused adult human material with rabbit eggs stripped of their chromosomes and created rabbit-human hybrid embryos which developed to approximately the 100-cell stage, about four days of development. Moreover, the scientists claimed to derive from these embryos stem cells similar to conventional human embryonic stem cells.

Historic attempts at human-ape
There are well-documented reports that a few scientists in the mid-1920s made serious attempts to create a half-human, half-chimpanzee. One of the Soviet Union's top scientists, Professor Ilya Ivanov, tried to impregnate female chimpanzees with human sperm in Africa in order to create a human-chimpanzee hybrid (a humanzee). These experiments were unsuccessful, but at the time many colleagues believed it was probably feasible.

Genetic barriers – a helpful concept?

It may be said that any form of mixing violates natural boundaries – it breaks the species barrier. To pursue this, however, we need to understand the strengths and weaknesses of the concept of species boundaries. Although it is rare for species to interbreed, the 'barrier' is in reality difficult to define.

> **It may be said that any form of mixing violates natural boundaries – it breaks the species barrier**

First, if each species has a clearly defined genome, then mixing species means mixing up two distinct genomes. But with the human genome, things are not that clear. To start with, around half the genes in human cells create proteins that keep cells alive and growing. These genes are found in many different living organisms where they vary only slightly, if at all, from the versions found in humans. This is why people quote figures such as 'humans are 50% banana'. It is therefore difficult to describe these so-called 'housekeeper' genes as belonging to any particular species.

Secondly, the human genome carries many genes that have no known function in humans, but are known to have specific roles in other animals. The human genome, for example, carries the entire gene sequence for the mouse tail; the cells simply miss the switch to turn it on. Some people therefore argue that adding more mouse genes to a human cell would not be doing anything new, though of course there would be the intention of introducing a new structure or function. In addition, retroviruses constantly carry new genetic material across species into chromosomes. A careful analysis of any organism shows that these viruses have been frequent visitors throughout generations.

Another, more intriguing, view of human beings sees us as communities of organisms. Each of us carries around 100 trillion micro-organisms that live primarily on our skin and in our guts. One paper estimates that humans carry more than 500 different species of micro-organism, and that together this means we carry 100 times as many genes as are found in our 'own' cells.

A further argument used against mixing individuals is that it will violate their genetic uniqueness. That, again, is not as clear as it might seem, because same-species chimeras are probably quite frequent in nature. Some will occur when two embryos fuse as they grow in the womb (see box), but other 'microchimeras' are created when cells from the fetus and placenta break off during pregnancy and birth and enter the mother's blood stream. Colonies of these cells may persist for decades, and on occasions these cells have found their way across the placentas in future pregnancies and become part of the makeup of the bodies of subsequent siblings. Some estimates claim that up to 50% of women who have been pregnant will be chimeric.

Arguments from 'nature'

From all these points, it is difficult to argue against hybrids or chimeras on a purely genetic basis. The issue then becomes less the actual composition of individual people's genomes, but how that composition came into being. Does the simple fact that something occurs in nature give us permission to do the same in the laboratory, and extend it further?

We need to be careful of falling into the trap of assuming that if something occurs in 'nature' then it must be good. Nature presents plenty of examples of actions that seem undesirable, ranging from disease to earthquakes. Similarly, medicine is a discipline that aims to fight off the worst effects of natural actions – if nature really shows us the way, then medicine should be confined to helping people who have physical injuries.

Undermining human dignity

Some people worry that to produce creatures that blur the boundaries between humans and animals could threaten to undermine the concept of human dignity since it is a dignity specifically reserved to humankind. Moreover, other commentators suggest that we should prevent future ethical dilemmas by forbidding anyone from trying to create an animal that may to some extent exhibit human capacities.

⇨ The above information is an extract from a document published by the Christian Medical Foundation and is reprinted with permission. Visit www.cmf.org.uk for more information or to view the full text and references.

© CMF 2007

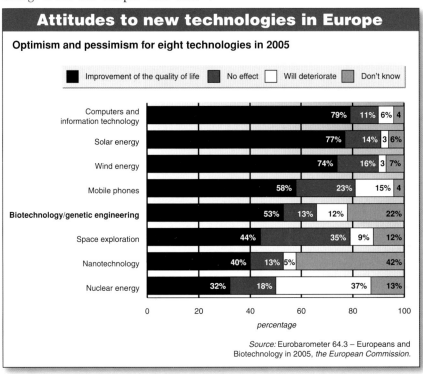

Attitudes to new technologies in Europe

Optimism and pessimism for eight technologies in 2005

Legend: Improvement of the quality of life | No effect | Will deteriorate | Don't know

Technology	Improvement of the quality of life	No effect	Will deteriorate	Don't know
Computers and information technology	79%	11%	6%	4
Solar energy	77%	14%	3	6%
Wind energy	74%	16%	3	7%
Mobile phones	58%	23%	15%	4
Biotechnology/genetic engineering	53%	13%	12%	22%
Space exploration	44%	35%	9%	12%
Nanotechnology	40%	13%	5%	42%
Nuclear energy	32%	18%	37%	13%

percentage

Source: Eurobarometer 64.3 – Europeans and Biotechnology in 2005, the European Commission.

Animal-human hybrids: it makes sense to say no

Information from CORE

A group of UK scientists held a press conference in London this week, expressing rage that their licence applications, asking to be allowed to clone embryos combining animal eggs with human tissue, were unlikely to be granted.

Readers of The Times, London, would be under the impression that no one but religious groups and opponents of embryo research are opposed to this cross species cloning.

This is definitely not the case and many countries have already banned this research, including France, Germany, Italy, Canada, and most recently Australia. Concerns address ethical and safety issues of great complexity and all these countries have come down firmly against animal-human hybrids.

For your interest we are collecting comments by international scientists, seriously concerned with responsible science, and who do not endorse the animal-human route as a way to investigate or cure human disease. We quote some of these and would like to point out that the authors are neither pro-life nor religious.

It is lamentable that a journal as prestigious as The Times has presented such a polarised position to its readers, and ignored the reality of the debate in question. The false hopes extended to patient advocacy groups are a tragic outcome of such inadequate journalism.

Stuart Newman, Professor of Cell Biology, New York Medical College, Founding Member of the Council for Responsible Genetics:

'I think this is a poorly motivated experiment.'

'Growth and development of the human-cow hybrid clone would say very little about the potential of a human-only clone to develop in the same fashion.'

Dr William James Peacock, Australian Chief Scientist, Chairman of National Science Forum, Molecular biologist, and acknowledged cloning supporter:

'Stem cell research could proceed without using animal eggs as incubators.'

'The use of animal eggs is not necessary.'

In December 2006 the Australian Senate voted without dissent to maintain a prohibition on human-animal hybrids.

Human Genetics Alert:

'The research in question is unnecessary and unlikely to provide useful information.'

'The use of animal eggs to produce stem cells seems designed to produce as many gene expression artefacts as possible.'

'Any conclusions regarding the differentiation of stem cells would be highly suspect, and such cells could not be used in treatment since their properties are unknown.'

'There is also the possibility of contamination by animal viruses.'

Royal Society:

'There is at present insufficient scientific justification for creating human-animal hybrid embryos.'

Professor Sir John Gurdon, Cambridge University uses similar technology to in-vestigate how eggs appear to be capable of converting adult cells into stem cells that can potentially grow into any tissue in the body. His experiments have focussed on injecting human DNA into frog eggs:

'Scientifically, I'm not persuaded it will work. If you put cells from one species into the egg of another, the egg may divide, but you could get a lot of genetic abnormality that won't lead to good quality stem cells.'
5 January 2007

⇨ The above information is reprinted with kind permission from CORE. Visit www.corethics.org for more information.

© CORE

Hybrid embryo ban 'would cost patients' lives'

Hundreds of thousands of patients with diseases of the nervous system will miss out on potentially life-saving new treatments if regulators ban experiments using part-human, part-animal embryos, scientists said yesterday.

The Human Fertilisation and Embryology Authority (HFEA) is next week expected to turn down applications from two teams of British researchers to transfer human cells into rabbit, cow and goat eggs.

Patients with diseases of the nervous system will miss out on potentially life-saving new treatments if regulators ban experiments using part-human, part-animal embryos

Scientists want to create the hybrid embryos that would be around 99.9 per cent human and 0.1 per cent animal in order to produce embryonic stem cells – the body's basic building blocks that can grow into all other types of cells.

They hope to use stem cells to both understand and provide new treatments for diseases such as Alzheimer's, Parkinson's, cystic fibrosis, motor neurone disease and Huntington's.

Until now the process of creating an early embryo by putting human DNA into an egg that has its nucleus removed – known as therapeutic cloning – has been carried out using human eggs from consenting IVF patients. However, these are in short supply and success rates have been low.

Chinese scientists have shown it is possible to harvest stem cells

By Nic Fleming, Science Correspondent

from embryos created by transferring human cells into rabbit or cow eggs.

The UK-based researchers, led by Dr Stephen Minger at King's College, London, and Dr Lyle Armstrong at the North East England Stem Cell Institute, in Newcastle, stress the hybrid embryos would be destroyed by 14 days when they are no bigger than a pinhead.

Last month the Government published a White Paper that will form the basis for an overhaul of laws on fertility treatment and embryo research.

It included a proposed ban on the creation of embryos that are part-human, part-animal, with a provision to allow such research in certain conditions under licence.

Dr Minger said yesterday: 'Informally we have been told [by the HFEA] they are unlikely to grant permission for our applications.

'At present we have no therapies to even alleviate the symptoms for conditions such as Alzheimer's, spinal muscular dystrophy and motor neurone disease, never mind make an impact on disease progression.'

Prof Chris Shaw, a neurologist from King's College London, said: 'I think this technique has the potential for very important outcomes for patients. To shut this research down at the moment would be an affront to those patients.'

The authority board will issue a policy statement that will inform its decision on the research licence applications at a meeting on Wednesday.

Opponents have described the proposed research as undermining 'the whole distinction between animals and humans'.

Josephine Quintavalle, the director of Comment on Reproductive Ethics, said: 'This kind of research makes people feel uncomfortable. There has been a groundswell of public concern and I think the HFEA has realised that.'

5 January 2007

© Telegraph Group Limited, London 2007

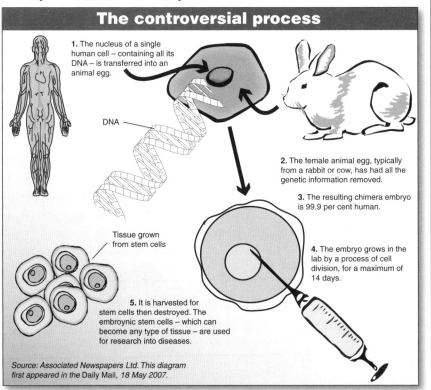

The controversial process

1. The nucleus of a single human cell – containing all its DNA – is transferred into an animal egg.

DNA

2. The female animal egg, typically from a rabbit or cow, has had all the genetic information removed.

3. The resulting chimera embryo is 99.9 per cent human.

4. The embryo grows in the lab by a process of cell division, for a maximum of 14 days.

Tissue grown from stem cells

5. It is harvested for stem cells then destroyed. The embroynic stem cells – which can become any type of tissue – are used for research into diseases.

Source: Associated Newspapers Ltd. This diagram first appeared in the Daily Mail, 18 May 2007.

Stem cells and human embryos

Medical science FAQs

Stem cells and human embryos are often mentioned together in the media, but how are they connected?
There are two sources of stem cell: adult and embryonic. Adult stem cells are found in bone marrow, muscle, skin, the brain and the digestive tract. These cells sit dormant in tissue until disease or tissue injury sparks them into action to repair damage. In this sense, scientists think that adult stem cells are restricted to maintaining the health of the tissue where they are found. In contrast, embryonic stem cells have the potential to turn into any cell type: if we can harness and influence their adaptability, they might be a source of healthy tissue to replace that which is diseased or damaged in adults.

The process of removing embryonic stem cells is regulated by the Human Fertilisation and Embryology Act (1990) which was amended in 2001 to include removing stem cells from embryos. The embryos from which these cells are taken are those created but not used for in vitro fertilisation treatment (known as IVF) and where the parents have consented. The cells are removed at the 5-day stage when the embryo is still a ball of cells, has not developed a nervous system and does not resemble a foetus.

The stem cells taken from such embryos have the potential to turn into any tissue type but cannot develop into a person. Researchers would like to continue studying the potential of these embryonic stem cells to provide treatments for Parkinson's disease, spinal injury and many other currently incurable conditions. It is because embryonic stem cells are obtained by removing them from an embryo that embryos and stem cells are often mentioned together in the media.

What is the difference between reproductive and therapeutic cloning?
The aim of therapeutic cloning is to create embryonic stem cells for use in medical research and development of stem cell therapies. In contrast, the aim of reproductive cloning is to make a fully developed replica of the organism being cloned. Human reproductive cloning is illegal in the UK and the practice is considered to be unethical.

For the purposes of human therapeutic cloning, the clone is made by inserting genetic material (nucleus) from the donor (for example from a human skin cell) into a human egg cell that has had its own nucleus removed. This process is called cell nuclear replacement. A current of electricity is applied to stimulate the egg cell to divide like an embryo. After 5-6 days, cells are taken out of the embryo to provide embryonic stem cells. It is illegal for the embryo to develop beyond 14 days, when the primitive streak, the early nervous system, appears.

Reproductive cloning also uses cell nuclear transfer. It was originally developed to improve animal breeding and is the process that created Dolly the Sheep, the first cloned mammal. The cloned Dolly embryo was implanted into the womb of a sheep and allowed to develop to full-term.

Scientists hope that, in the long-term, the process of therapeutic cloning will yield embryonic stem cells that they can then use to develop replacement tissue to treat conditions like diabetes and motor neurone disease. Other applications of embryonic stem cell research are in increasing knowledge about infertility, miscarriage, contraception, genetic abnormalities, embryo development and serious disease.

⇨ The above information is re-printed with kind permission from the Association of Medical Research Charities. Visit www.amrc.org.uk for more information.
© AMRC

Stem cell research: new horizons in medicine

In the late 1990s, American scientists developed a method of culturing embroynic stem cells – cells with the potential to develop into any of the specialised tissues the body needs. This raised hopes of a major emdical breakthrough – the ability to replace diseased or worn out body parts with new, healthy tissue. Although there are still many technical and ethical issues to be resolved, the new techniues have the potential to revolutionise medicine.

Therapeutic stem cell cloning

In this technique, DNA is taken from the cells of a person and used to produce embryonic syem cells, which may then be differentiated into new tissues or organs. The hope is that this technology will help people who are seriously ill with problems ranging from diabetes and Parkinson's disease to heart attacks and spinal injuries. The technique still needs a lot of development, but its medical potential is enormous.

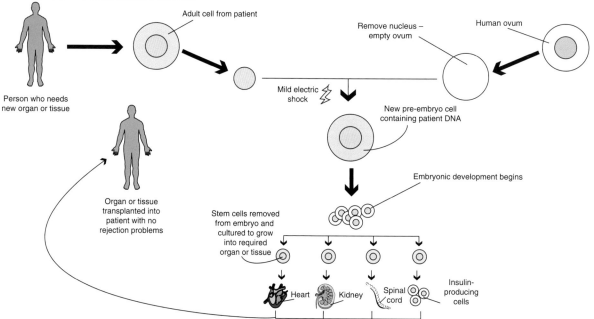

Embryonic stem cells

The early human embryo contains many stem cells which can be harvested and culture in the laboratory to produce huge numbers of undifferentiated cells. By changing the culture conditions scientists have persuaded some of the stem cells to differentiate into new tissues including cartilage, bone, nerve and intestine. Whole organs for transplants will be the ultimate challenge.

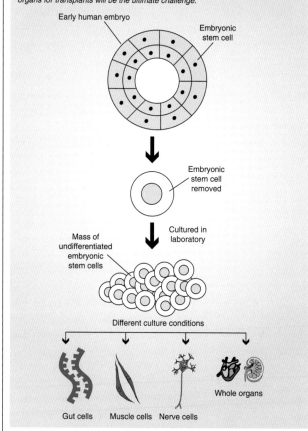

Adult stem cells

Adults have stem cells in their bodies – but not ver many of them. What's more, they can be fiddicult to extract and to use, and most can only form a limited range of different cell types. However, they are already used successfully in bone marrow transplants and there is some promising work on using them to repair damaged heart muscle and to reverse Parkinson's disease. Interest in the future of this technology is growing steadily.

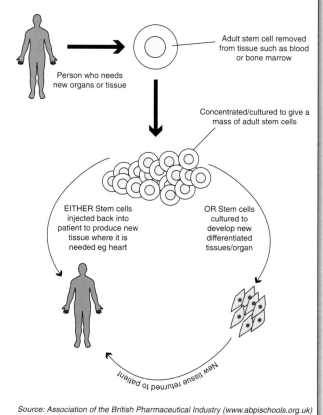

Source: Association of the British Pharmaceutical Industry (www.abpischools.org.uk)

Issues

www.independence.co.uk

16

Frequently asked questions on stem cells

Information from the North East England Stem Cell Institute

What is DNA?

All human cells contain 23 pairs of chromosomes, one set inherited from each parent. Each pair of chromosomes are made up of a long strand of DNA, which contains individual genes. Taken together, this genetic material of genes and chromosomes provides a blueprint or plan for the cell itself and for the body in which it lives.

What is Mitochondrial DNA?

The mitochondria are tiny structures contained within cells that provide the energy to cells, like batteries. Any cell may contain thousands of mitochondria, each of which has its own DNA, containing only 37 genes which are used only to make the mitochondria itself work (they have no influence over the DNA of the cell, so cannot effect the body itself, for example by changing eye colour).

Why is stem cell research controversial?

There is very little controversy about using bone marrow stem cells or other adult (somatic) stem cells, such as those found in umbilical cord blood. However, for some people, the creation of stem cells from embryos is unethical. Many people believe that human life begins at conception so they have a strong moral view that research should not use embryos. In the US, for example, research on embryonic stem cells cannot at present be funded by the Federal government – although some states, such as California, do fund this work and it is possible that the law will change in the future.

Do embryos have to be destroyed to make embryonic stem cells?

The embryonic stem cells are derived at an early stage of human development called the blastocyst stage about five days after fertilisation. This is before the embryo would implant in the womb. The cells are found in an area of the blastocyst called the 'inner cell mass', which contains the cells which may go on to form a fetus at a later date. The blastocysts that are used in this process are donated by couples who are undergoing fertility treatment and have volunteered to donate fertilised eggs that are not going to be placed back in the woman's body. They would be discarded if not used for research.

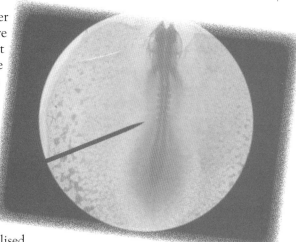

Chicken embryo

For some people, the creation of stem cells from embryos is unethical. Many people believe that human life begins at conception

---

Who governs this research and treatment?

All medical and scientific research in the UK is strictly governed by a variety of bodies, including the local ethics committee (www.corec.gov.uk) and the Department of Health Research Governance Framework for Health and Social Care, (www.dh.gov.uk). In the case of eggs and embryos donated for a research project that is subject to legal regulation, a licence is required from the Human Fertilisation and Embryology Authority (www.hfea.gov.uk). New treatments are introduced in the UK only after appropriate research and under the principals of good medical practice. This ensures that patients are given full details of the proposed treatment including the risks and success rates. The Medicines and Healthcare Regulatory Authority regulates the production and use of products that might be derived from stem cells for use in treatments. Stem cells and any cells derived from them that will be used directly for treatment will be regulated by a new authority to be established under the EU Tissue Bank Directive.

Can stem cells be turned into any type of cell?

Embryonic stem cells can and we are working on finding out how this happens. We also think that certain very rare types of stem cell found in cord blood can turn into any cell type. Adult stem cells are designed to produce one specific cell type only, but we are working to see how they can be re-programmed to maybe produce other types of cell.

Have stem cells been used to treat any human diseases yet?

Yes. They have been used in bone marrow transplants for a long time now. They are used to treat diseases of the blood, such as leukaemia. There have been some media reports that other stem cell treatments have taken place in various places. However, most scientists are sceptical of these claims and would wish to see more detailed clinical trials take place before they can be certain that the treatments claimed have actually worked.

When will stem cells be used with patients?

Clinical trials, using stem cells, which involve checking very carefully whether new medicines or treatments actually work, are starting in many countries at the moment. We expect that some of these will prove effective and that we may see their common use within the next few years or so. It is likely that the first therapies will use adult stem cells and that treatments based on embryonic stem cells will be further off.

How can stem cells be used to help people?

We hope that it will be possible to use stem cells to regenerate human tissues. In normal human bodies, stem cells work by renewing the cells, but sometimes this does not work for a range of reasons. Stem cell therapies would be based on using the power of stem cells either to work better at repairing and regenerating the cells they are designed to help or by re-programming them to make different types of tissue which, for whatever reason, is either damaged or the body isn't repairing naturally.

Which diseases will stem cells cure?

Potentially, any diseases which prevent the body naturally renewing itself or which have destroyed tissue are targets for stem cell-based therapies. The most likely targets at this stage – and those we are concentrating on in the North East – seem to be diabetes; diseases of the blood and liver; degenerative diseases of the neuronal system, such as Parkinson's disease; and injuries or damage to skin and eyes.

Could stem cells have saved superman?

Christopher Reeve, the actor who played superman, was severely disabled in a horse riding accident some years ago. He became a passionate supporter of stem cell research, believing that it could lead to ways of generating neuronal cells, which could be used in the repair of spinal cords. Sadly, he died before the science could progress to this point, but it is certainly one of the possible outcomes that we are pursuing for the future.

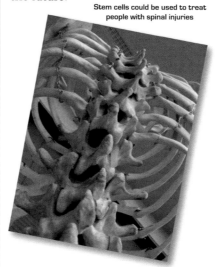
Stem cells could be used to treat people with spinal injuries

Who will decide when stem cells can be put in patients?

Medical research in the UK is strictly governed by law, a variety of national bodies, and local ethics committees who review every research proposal at a local level. The process involves strict control over the basic scientific research, then moving on to trials in animals, trials with humans (clinical trials) and, finally, following extensive review, transfer into hospitals for general use.

What is a stem cell 'line'?

Once stem cells have been isolated, they are cultured, ie grown up using special equipment, so that there are enough of them to do experiments on. A group of these cells is called a 'line'.

What is cloning?

Some scientists have worked on a procedure known as nuclear transfer, which involves transferring the genetic material from one animal into the egg of another. If the egg is developed into an embryo and then grown, it will become identical to the donor. This process has been carried out in a number of animal species, including mice and dogs. The most famous example, however, was Dolly the sheep, who was successfully cloned in Edinburgh in 1996.

Can I create a clone of myself?

No! Cloning in humans is illegal in the UK and elsewhere. Some scientists have worked on human nuclear transfer, including a team in Newcastle who were the first to create a cloned blastocyst, announced in May 2005. But this therapeutic cloning can only be for experiments: reproductive cloning is not allowed and any embryo created by this process would not be allowed to develop beyond 14 days.

How can stem cells be used to develop drugs?

If stem cell lines were created in laboratories, they could look and act like human tissues. Rather than experimenting on animals, it might be possible to use these stem cells to test drugs on.

Who is leading the research into stem cells?

Scientists throughout the world are very interested in this work and new teams are being formed all the time. There are particular research strengths in South East Asia, the US and some European countries, particularly Britain. The UK has some real strengths in this field, with maybe half a dozen or so centres of excellence, including the North East of England. We are keen both to further develop our research and to link up with other scientists in the UK and beyond who are interested in working with us.

How can I help aid the research?

The most important thing is to understand the issues and talk to other people about them. We want to involve the public in our research as much as possible. We also have a charitable fund which helps our (often very expensive) research – if you can help raise money, please let us know. Finally, we are often asked by people if they can take part in clinical trials. Although we are not yet at this stage, we will of course in the future be looking for people to take part.

⇨ The above information is re-printed with kind permission from the North East England Stem Cell Institute. Visit www.nesci.ac.uk for more information.

© NESCI

Statistics from *Eurobarometer 64.3 – Europeans and Biotechnology in 2005: Patterns and Trends.*

Familiarity with stem cell research across Europe

Country	Very familiar	Fairly familiar	Not very familiar	Not at all familiar
EU	4	26%	37%	32%
Denmark	14%	47%	30%	10%
Italy	10%	38%	35%	16%
UK	5%	40%	35%	20%
Hungary	7%	33%	36%	24%
Netherlands	3	37%	41%	19%
Luxembourg	9%	27%	36%	29%
Spain	6%	27%	38%	28%
Ireland	5%	29%	33%	34%
Belgium	5%	28%	34%	32%
Sweden	4	27%	48%	21%
Finland	3	23%	43%	30%
France	2	24%	28%	47%
Austria	4	20%	44%	32%
Portugal	3	20%	35%	42%
Cyprus	4	18%	35%	43%
Germany	3	17%	43%	37%
Poland	2	17%	39%	42%
Malta	2	15%	35%	49%
Slovenia	1	14%	48%	38%
Slovakia		13%	56%	30%
Czech Republic		12%	47%	40%
Estonia	1	11%	30%	58%
Latvia	1	9%	31%	59%
Greece	2	7%	24%	67%
Lithuania	1	8%	39%	52%

% 0 20 40 60 80 100

European views on embryonic and non-embryonic stem cell research

Legend: Embryos / Umbilical cords

	Embryos	Umbilical cords
Approve with usual government regulation	23%	28%
Approve with tighter regulation	36%	37%
Do not approve except under special circumstances	17%	14%
Do not approve under any circumstances	9%	7%
Don't know	15%	14%

Ethical issues around stem cell research

Legend: Agree / Disagree / Don't know

	Agree	Disagree	Don't know
Wrong to use human embryos in medical research	41%	41%	18%
Embryonic stem cell research should be allowed	53%	29%	18%
Science should prevail over ethics in stem cell research	53%	29%	19%

% 0 20 40 60 80 100

Frequency of religious service attendance and approval of stem cell research

Apart from weddings or funerals, about how often do you attend religious services?

- Once a week or more
- Between once a month and once in three months
- Only on special holy days/about once a year
- Less often than once a year/never

	Once a week or more	Once a month–3 months	Special holy days	Less often/never
Approve with usual government regulation	26%	36%	45%	56%
Approve if more tightly regulated	63%	76%	82%	70%
Do not approve except under very special circumstances	36%	41%	36%	31%
Do not approve under any circumstance	32%	16%	15%	14%
Don't know	43%	31%	23%	28%

Overall approval of embryonic stem cell research

What information would people want about embryonic stem cell research?

Legend: Mentioned first / Mentioned second

	Mentioned first	Mentioned second
Benefits and risks?	39%	20%
Current regulations and regulation enforcers?	13%	21%
Who is responsible for moral limits?	14%	17%
What processes and techniques used?	12%	16%
Who are funders and beneficiaries?	8%	11%

Source: Eurobarometer 64.3 – Europeans and Biotechnology in 2005, *the European Commission.*

Genetic engineering

Should genetic engineering become more commonplace?

A positive 59% of Britons agree with controversial stem cell research, according to research by ICM conducted on behalf of the Guardian. The slight majority of respondents, 51% are in support of genetic engineering when designed to correct physical defects in unborn children. This indicates a positive change in the public's attitude to genetic engineering and is good news for the scientists who are adamant that cloning cells is the way to find cures for some of our most debilitating genetic diseases.

Stem cell research is advancing apace. Stem cells are cells that are capable of growing into any of the 300 different kinds of cell in the human body and researchers currently extract them from human embryos that have been discarded during fertility treatments.

However, British scientists are currently seeking approval to create embryos by fusing human cells with animal eggs in controversial research, which will boost stem cell science and tackle some of the most debilitating and untreatable neurological diseases. If they get approval from the Human Fertilisation and Embryology Authority they plan to create embryos that will be 99.9% human and 0.1% rabbit or cow by fusing human cells with animal eggs. They will then use the embryos to create stem cells that carry the genetic defects responsible for neurological conditions such as motor neurone disease. By converting the stem cells into neurons, the scientists will be able to unravel how the disease destroys nerves and identify drugs to stop or reverse the damage.

The HFEA is also currently undertaking a public consultation on human egg donation, which is gathering views on whether scientists should be allowed to seek altruistic donations of eggs for research. The ICM research seems to suggest that this might not be as controversial as we once thought.

By Antonia Windsor

However, although the public are largely on side on the use of genetic engineering to help cure illness or provide transplant organs, there is still an overwhelming resistance to parents creating 'designer babies'. Only 13% of respondents to the ICM survey said they were in support of the parent's right to use genetic engineering to design their unborn child, with 63% opposing the process. However, a higher 20% of 18-24 year olds said they would condone the practice, which suggests perhaps that future generations will be more open to the idea.

A positive 59% of Britons agree with controversial stem cell research

Currently it is only legally possible to carry out two types of advanced reproductive technologies on humans using In Vitrio Fertilisation to fertilise eggs with sperm in test tubes outside the mother's body. The first involves choosing the type of sperm that will fertilise an egg: this is used to determine the sex and the genes of the baby. The second technique screens embryos for a genetic disease: only selected embryos are implanted back into the mother's womb. This is called Pre-implantation Genetic Diagnosis.

In the future we may be able to 'cure' genetic diseases in embryos by replacing faulty sections of DNA with healthy DNA. This is called germ line therapy and is carried out on an egg, sperm or a tiny fertilised embryo. Such therapy has successfully been done on animal embryos but at present it is illegal to do this in humans.

Fears are that this technique will be taken a stage further and used to select personality traits in the unborn child, from their hair or eye colour, to their ability to perform well in sports or exams. However, the results of the survey show that it will be some time before public opinion turns in favour of us playing God with our unborn children, while the way seems clear for scientists to continue the advancement of genetic engineering to help us live longer, healthier lives.

October 2006

© *Guardian Newspapers Limited 2007*

Your stem cell body repair kit

Scientists believe stem cells could prove to be the ultimate body repair kit, with no need for donated organs, man-made joints or drugs to keep failing body parts working. So what are these potent sources of new life? They are cells at such an early stage of development that they can, in theory, be turned into any type of tissue. While associated with discarded IVF embryos and blood from newborn babies' cords, stem cells can also come from adult bone marrow, muscle and skin. Here, Pat Hagan reports on the latest in stem cell medicine.

Spine

The late superman actor Christopher Reeve campaigned for stem cell research after breaking his spine. Some research teams have cured animals of paralysis, using stem cells harvested from their bone marrow to encourage damaged nerves to regrow. Brazilian scientists even claim to have restored feeling to 12 paralysed patients treated with their own stem cells and given electrical stimulation.
AVAILABILITY: Five years. In UK, currently being tried on motorbike accident victims.

Pancreas

The pancreas is the body's insulin factory. We need insulin to help muscles absorb sugar from our diet and turn it into energy. In patients with Type 1 diabetes, the pancreas stops producing insulin, or does not make enough, and too much sugar circulates in the blood, eventually causing irreversible organ damage. Diabetics therefore have to inject themselves with insulin several times a day.

Scientists at Toronto University claim they have found adult stem cells in the pancreas that could be extracted, cultivated into millions of cells, and injected back into the body. The healthy new pancreas tissue would hopefully produce more insulin.
AVAILABILITY: Ten years at least.

Knees

Excess fat round the tummy contains stem cells that could cure injured knees.

Scientists in America recently discovered a protein that turns stem cells in belly fat into tissue resembling cartilage. It means patients with arthritis could one day have fat drained off their tummy to produce new cartilage cells, which would be injected back into joints.

British experts at Imperial College London have also succeeded in turning stem cells harvested from embryos into healthy new cartilage.
AVAILABILITY: 5-10 years.

Bladder

Urinary incontinence, which affects thousands of people in the UK, could be cured using adult stem cells extracted from muscle tissue in the arm.

These are cells that are already programmed to turn into muscle – so when they are injected into the neck of the bladder, they boost muscle tone and reduce leakage. Austrian doctors are already using the technique to treat women affected by the condition.
AVAILABILITY: Already used in Austria but not yet available in the UK.

Blood

Cancer specialists have been using a form of stem cell therapy for blood cancers for nearly 20 years. The technique, haematopoietic cell transplantation (HTC), is used on patients with leukaemia. HTC cells are extracted from donated bone marrow and turned into a whole range of different types of blood cells. These are then injected into the sick patient, where they produce healthy blood cells.
AVAILABILITY: Already used in the UK for leukaemia.

Face

Even cosmetic surgery could be transformed by cell therapy. An American firm, Isolagen, is already offering a treatment in the UK to banish wrinkles.

It involves taking a tiny skin sample to grow vast quantities of fibroblasts – the cells in the body that produce collagen. A fibroblast

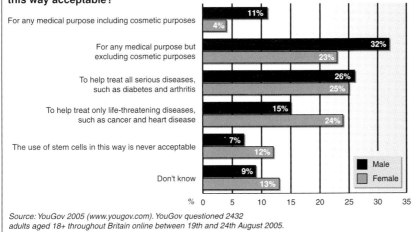

How should stem cells be used?

Respondents were asked 'Some scientists plan to use cloned early embryos as a source of "stem cells" – the flexible parent cells that produce every kind of tissue in the body. By growing embryo stem cells, these scientists hope to create unlimited supplies of tissue to treat diseases such as diabetes, Alzheimers' disease and heart disease. In your view, when, if ever, is the use of stem cells in this way acceptable?'

	Male	Female
For any medical purpose including cosmetic purposes	11%	4%
For any medical purpose but excluding cosmetic purposes	32%	23%
To help treat all serious diseases, such as diabetes and arthritis	26%	25%
To help treat only life-threatening diseases, such as cancer and heart disease	15%	24%
The use of stem cells in this way is never acceptable	7%	12%
Don't know	9%	13%

Source: YouGov 2005 (www.yougov.com). YouGov questioned 2432 adults aged 18+ throughout Britain online between 19th and 24th August 2005.

is effectively a stem cell that ahs already matured into a specific type of cell; one that looks after the body's collagen needs.

Collagen is the body's 'scaffolding', keeping skin taut and wrinkle-free. The cells are injected back into the face where they make new collagen.

AVAILABILITY: Now – at a cost of £3,000 to £4,000. Go to www. isolagen.co.uk

Brain

Stroke victims are often left partially paralysed because of damage to their brains. At Stanford University in the US, tests on mice revealed that stem cells injected near the site of the damage quickly turned into brain cells, called neurons.

Neurons help transmit messages in the brain and to the rest of the body – something which is often disrupted after a stroke.

Now research is underway to see if this breakthrough actually improves brain function in stroke patients.

AVAILABILITY: At least five years.

Eyes

Age-related macular degeneration is the leading cause of blindness in Britain, affecting around 250,000 people. Yet scientists hope it could one day be cured by stem cells harvested from the back of the eye.

British and Canadian experts found the cells were able to regenerate into healthy photoreceptor cells that help to process images for the brain. In AMD, these cells become damaged beyond repair.

AVAILABILITY: Five to ten years.

Hair

Scientists at the University of Pennsylvania working with mice have managed to grow hair in the lab using adult stem cells extracted from hair follicles. They hope the same technique will work in humans.

Meanwhile, British company Intercytex is testing a similar technique that doesn't actually use stem cells. Instead, it involves taking a tiny sample of skin from the head and using it to mass-produce dermal papilla – the cells which have already matured into hair-growing factories. These are then injected back into the scalp potentially giving a full head of hair.

Existing hair transplants rely on transferring individual hair follicles rather than millions of active cells.

AVAILABILITY: Between three and five years.

Ears

A team at Sheffield University is exploring whether stem cell from embryos could restore hearing to the deaf. They are developing ways of turning these 'raw' cells into ones that will replace sensory hair cells lost or damaged through ageing, loud noise or illness.

As sound waves travel into the ear, the eardrum vibrates, disturbing fluid in the inner ear. This stimulates millions of tiny hairs that convert the movement into electrical impulses and dispatch them to the brain.

Once these hairs are destroyed, they do not regenerate themselves.

Dr Marcelo Rivolta, a research scientist at Sheffield, claims stem cells can be used to grow new hairs in a damaged ear.

AVAILABILITY: Ten years.

Heart

British patient Ian Rosenberg made headlines in 2004 when his life was transformed after stem cell treatment in Germany for heart disease. The treatment involves taking stem cells from the bone marrow in a patient's leg and injecting them directly into the heart, where they turn into healthy new muscle tissue. Mr Rosenberg now runs the Heart Cells Foundation to help more patients access the treatment. Go to www. heartcellsfoundation.com

AVAILABILITY: Now – only privately at the Johann Goethe University Hospital in Frankfurt. A four-year UK trial involving 700 patients is underway.

Liver

Trials are underway ay Hammersmith Hospital in London using adult stem cells extracted from the patient's own bone marrow to treat a damaged liver.

Laboratory tests show the cells boost liver function by replacing those lost through cirrhosis. As healthy new liver tissue grows, the organ's ability to cope with toxins such as alcohol increases.

AVAILABILITY: Five years at least.

⇨ This article first appeared in the *Daily Mail*, 29 August 2006

© 2007 Associated Newspapers Ltd

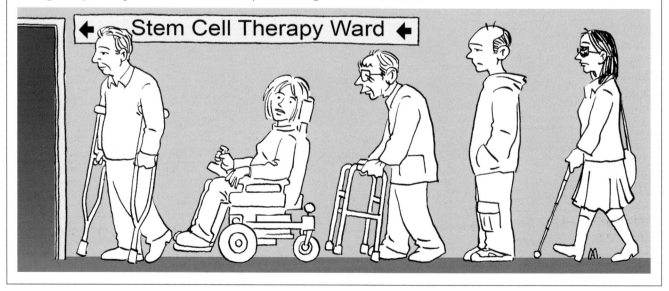

Stem cells – too fast too soon?

The hottest topic in biomedical science may be poised to leap from lab to clinic. Stem cell therapy seems almost too good to be true: the chance to mend or replace damaged body tissue. Is it time to try it on people or are we in danger of getting carried away?

By Lisa Melton

Stem cells, some say, are the future of medicine. They have the potential to treat and reverse many of humanity's ills. Instead of drugs or the surgeon's scalpel, stem cells could become the body's own 'repair kit'. The prospects are hugely exciting. But as stem cell researchers accelerate towards developing clinical applications, some experts urge caution. Do we know enough about stem cell behaviour – and do we fully understand the risks?

The concept of stem cell therapy sounds simple enough. Put the stem cells in the right place in the body, and they will produce new cells and renovate the tissue. At least in theory, stem cells can be converted into any human tissue: insulin-secreting cells to treat diabetes, dopamine-producing neurons for Parkinson's disease, skin cells for burn victims, or heart cells to help repair the damage of a heart attack.

Heart help

In one field, cardiology, doctors are forging ahead. People who have had heart attacks are being given their own 'adult' stem cells (see Stem cell lexicon), which are taken from their bone marrow and injected into a vessel feeding the heart. Ten human trials of the marrow-to-heart approach have been completed in clinics around the world, all but one with positive results.

A medical success story? Not everyone is convinced. 'I think the risk is very high,' counters Jürgen Hescheler at the University of Cologne in Germany, a stem cell scientist and renowned critic of the trials. 'It's too early to go into humans.' Other researchers agree. 'Stem cell therapy does have a future, but nobody benefits by blazing a trail early on,' argues Michael Marber, a cardiologist at St Thomas' Hospital in London. 'It artificially raises people's hopes.'

> **Stem cells, some say, are the future of medicine. They have the potential to treat and reverse many of humanity's ills**

Better therapies for heart disease are certainly needed – the condition is one of the West's biggest killers, accounting for about one in five deaths. A heart attack occurs when one of the organ's blood vessels becomes blocked. The muscle supplied by that vessel is starved of blood and dies, causing the heart to lose pumping power. Medicines help, but only to a point; once cardiac muscle dies, it cannot be renewed. At least, that used to be the thinking.

Then, in 2000, a ground-breaking experiment in mice suggested that adult stem cells from bone marrow could develop into heart muscle [Orlic D et al. Bone marrow cells regenerate infarcted myocardium. Nature 2004;410(6829):701-5]. Excitement bordering on euphoria rippled through the cardiology community. This opened up the possibility of treating the disease with the body's own cells. In the mice the damaged heart was rebuilt by a spectacular 68 per cent and heart function improved dramatically. Would it work as well in humans?

Cardiologists wasted no time. Within about a year, three randomised clinical trials were launched in Germany. The results were hailed as a breakthrough. On average, 40 patients who received the stem cell therapy in a Frankfurt-based study improved their heart's pumping capacity by 8 per cent compared with 4 per cent in an untreated group. This may not seem much, but it could give someone the ability to walk up a flight of stairs, for example.

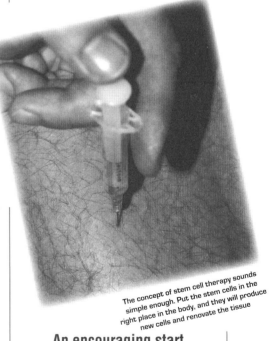

The concept of stem cell therapy sounds simple enough. Put the stem cells in the right place in the body, and they will produce new cells and renovate the tissue

An encouraging start

Unfortunately, the studies were not placebo-controlled, and, overall, the gains were modest. 'People are going ahead saying this is safe,' says Dr Hescheler. 'For me this is not proven.' There is an uncomfortable possibility that the bone marrow cells injected into the heart could develop into other tissues, such as bone, skin or even tumours.

So far, these problems have not surfaced. Yet scientists doing basic stem cell work are calling for a deeper understanding of what drives stem cells before rushing to the clinic. They fear that if people start being harmed it could set back the field for many years.

On the other hand, there is a precedent for cell therapy: bone marrow transplantation. 'Transplanting bone marrow is well established in clinical practice for blood restoration,' says Roger Pedersen, a professor of regenerative medicine and convenor of the Cambridge Stem Cell Initiative. Indeed, bone marrow transplants have a good safety record and have been used in clinical practice for decades.

> **Despite the uncertainties still hanging over the technique, stem cell transplantation 'clinics' have sprung up at sites around the world**

Even so, critical issues need resolving. 'What remains controversial is [the technique of] putting cells into specific organs. I think it remains to be demonstrated to be effective,' Professor Pederson notes. 'The response could be indirect, caused by a growth factor. If so, the same response could be accomplished by a purified protein without subjecting patients to the risks of cellular transplants.'

In the UK, clinical research is progressing. 'This is a potentially revolutionary treatment. It could transform the lives of people living with heart disease,' says cardiologist Anthony Mathur at Barts and the London Hospital. What is needed now, he insists, is a large-scale study to establish once and for all whether the treatment works. With cardiologist John Martin, at University College London, Dr Mathur is setting up a four-year research programme, the first of its kind to tackle this question. It will be a 600-patient blinded, placebo-controlled trial run in London.

Supply and demand

Despite the uncertainties still hanging over the technique, stem cell transplantation 'clinics' have sprung up at sites around the world, including Barbados, the Dominican Republic, Bulgaria and India. The Barbados operation is primarily a centre for 'cosmetic' stem cell therapy. Clients receive foetal stem cell implants that supposedly have a rejuvenating effect. How this effect might be achieved, the head of the centre acknowledges, is not known.

More seriously, for people with incurable conditions, stem cells mean hope where none existed before. This understandably creates great pressures to begin human treatments as rapidly as possible. Perhaps the furthest advanced is Beijing neurosurgeon Hongyun Huang from Chaoyang Hospital, who uses cells from aborted fetuses to treat patients with spinal injuries and degenerative disorders such as multiple sclerosis, amyotrophic lateral sclerosis and Parkinson's disease. At US$20 000 a go, it is not cheap, yet people are queuing up for treatment.

Dr Huang's critics argue, however, that his claims of success ride on anecdotal evidence, with little scientific backing. He has published in Chinese journals but not, so far, in peer-reviewed Western publications, which would mean his work could be fully scrutinised by the research community. 'Before we carry on, we really need to know, does this really work and what is likely to be the mechanism,' says Stephen Minger, a cell biologist at King's College London, who will also be involved in the upcoming cardiology trials in London. Never-theless, Dr Huang remains adamant that the human therapies he is pioneering will work and will come to be accepted.

Commercial push

In the West, companies are eager to move to the clinic. On the basis of its work on animals, Californian biotech company Geron argues that human interventions are likely be effective and safe for people with spinal cord injuries. They hope to start clinical trials by summer 2006, injecting foetal cells into patients with paralysing spinal injuries.

The spinal injury trials will be based on work led by Hans Keirstead, a stem cell researcher at the University of California, Irvine. He restored some mobility to rats with spinal injuries after injecting cells derived from human embryonic stem cells. However, many scientists view Geron's plans with scepticism, concerned that negative results could damage the field.

The UK-based company Re-Neuron is also using foetal stem cells, which it hopes to turn into a stable, off-the-shelf product, with all the practical advantages of a pharmaceutical drug.

'It's a product suitable for all

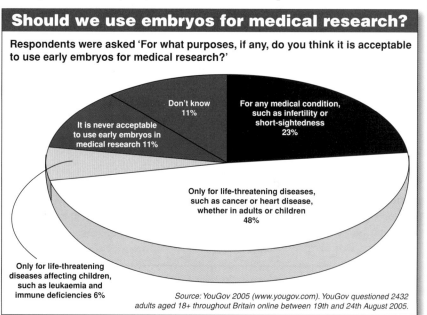

Should we use embryos for medical research?

Respondents were asked 'For what purposes, if any, do you think it is acceptable to use early embryos for medical research?'

- Don't know 11%
- It is never acceptable to use early embryos in medical research 11%
- For any medical condition, such as infertility or short-sightedness 23%
- Only for life-threatening diseases, such as cancer or heart disease, whether in adults or children 48%
- Only for life-threatening diseases affecting children, such as leukaemia and immune deficiencies 6%

Source: YouGov 2005 (www.yougov.com). YouGov questioned 2432 adults aged 18+ throughout Britain online between 19th and 24th August 2005.

patients,' says ReNeuron's John Sinden. 'It is available for injecting into the patient as and when it is required.'

The company's scientists have genetically engineered these cells with the c-myc gene to create long-lasting stem cell lines, circumventing the need to rely on foetal tissue, which is in limited supply.

ReNeuron's main focus is the brain, says Dr Sinden. 'We have identified a cell line that is useful for stroke, and there is evidence it might be effective for Huntington's disease.' Reneuron is waiting for IND (investigational new drug) approval to start clinical trials in USA next year. 'We are confident that in 12 months we could be in the clinic.'

Nevertheless, the use of genetically modified cell lines has raised concerns. The cells might grow unchecked, giving rise to tumours. ReNeuron has tracked transplanted mice for six months without detecting any such problems and is now testing its cell line in non-human primates (in the USA) to address this question.

Embryonic science

Even more versatile than adult stem cells are embryonic stem (ES) cells. Taken from the very early-stage embryo, these cells are a 'blank slate' capable of developing into any of the body's 300 cell types. In principle, ES cells could be used directly in patients, or scientists could coax ES cells in culture to become a desired cell type before using them. This is not an easy task. Understanding the signals that normally instruct these cells to choose a particular pathway will take some time.

Even then, another big hurdle remains. The implanted cells would be recognised as 'foreign', triggering a rejection-like immune response. One solution is to create a bank of different lines, with patients being given the cells with the closest match to their own.

Another strategy, though, would be to create a personalised ES cell line. Thanks to nuclear transfer, a person's own genetic material could be transplanted into an egg that had had its own nucleus removed. ES cells harvested from an embryo grown from such an egg would be genetically identical to those of the patient. This technique, which has produced encouraging results in animal experiments, was supposedly achieved in humans by South Korean researcher Woo-Suk Hwang in May 2005.

Hwang claimed to have created 11 new human ES cell lines by replacing the genetic material of donated unfertilised eggs with that of skin cells from people with genetic diseases or spinal injuries. But with Hwang's human research now discredited, it is not certain that this can be done. ES cells also raise a host of ethical issues need to be carefully considered before plans are developed for routine medical use.

Perhaps not surprisingly, views differ as to how long it will take for ES cells to become standard therapy. There is understandable excitement at their enormous potential, and for people with currently incurable conditions they may seem like a final lifeline.

But are some clinicians being too hasty in promising a solution to those patients who have run out of options?

'The push to get ES cells into the clinic 'tomorrow' is coming mainly from the US,' says Dr Minger. 'In the UK, most of us feel that we are a long way from the clinic, perhaps ten to twenty years. It's important that we do this right – that we are convinced that the cells are safe and clinically effective. We are trying to dampen down the hype. People need to understand that this will take years.'

Stem cell lexicon

The term 'stem cell therapy' actually covers a range of possible approaches with different types of cell:

Adult stem cells: Progenitor cells that generate new cells needed on an ongoing basis by adults. Typically generate a restricted range of cells and occupy particular niches in the body. Often difficult to isolate. Difficult to distinguish 'true' stem cells and partially differentiated cells.

Precursor cells: Partially differentiated cells, able to divide and generate new specialist cells, but of a limited number of types.

Embryonic stem (ES) cells: Cells isolated from an early embryo, capable of giving rise to all the cell types of an adult.

Therapeutically cloned embryonic stem cells: ES cells generated following single cell nuclear transplantation. Cells generated will be genetically identical to donor.

Foetal stem cells: Stem cells collected from fetuses following terminations.

Cord blood stem cells: Cells from the embryo present in blood vessels in the umbilical cord, with the potential to form many different cell types. Could be banked at birth and used later in life.

'De-differentiated' stem cells: Adult cells modified to become like stem cells. Not currently available, but being researched.

3 May 2006

⇨ The above information is reprinted with kind permission from the Wellcome Trust. Visit www.wellcome.ac.uk for more information.

© *Wellcome Trust*

An ethical solution to stem cell controversy?

Scientists create embryonic stem cells without destroying embryos. Researchers are 'optimistic' that the process can one day work in human cells

By David Cameron, Whitehead Institute

Scientists have created embryonic stem cells in mice without destroying embryos in the process, potentially removing the major controversy over work in this field. Embryonic stem cells are special because they are pluripotent, meaning they can develop into virtually any kind of tissue type. They therefore offer the promise of customized cells for therapy.

The work, which appears in the June 6 online issue of Nature, was led by Rudolf Jaenisch, a member of the Whitehead Institute and a professor of biology at MIT. His colleagues on the work are from Whitehead, MIT, Massachusetts General Hospital, the Broad Institute of MIT and Harvard, and Harvard Medical School.

Somatic cell nuclear transfer ("therapeutic cloning") offers the hope of one day creating customized embryonic stem cells with a patient's own DNA. In this process, an individual's DNA would be placed into an egg, resulting in a blastocyst that houses a supply of stem cells. But to access these cells, researchers must destroy a viable embryo.

Now, Jaenisch and colleagues have demonstrated that embryonic stem cells can be created without eggs. By genetically manipulating mature skin cells taken from a mouse, the scientists transformed these cells back into a state identical to that of an embryonic stem cell. No eggs were used, and no embryos destroyed.

"These reprogrammed cells, by all criteria that we can apply, are indistinguishable from embryonic stem cells," says Jaenisch.

What's more, these reprogrammed skin cells can give rise to live mice, contributing to every kind of tissue type, and can even be transmitted via germ cells (sperm or eggs) to succeeding generations. "Germ line transmission is the final and definitive proof that these cells can do anything a traditionally derived embryonic stem cell can do," adds Jaenisch.

"But," he cautions, "these results are preliminary and proof of principle. It will be a while before we know if this can ever be done in humans. Human embryonic stem cells remain the gold standard for pluripotent cells and it is a necessity to continue studying embryonic stem cells through traditional means."

Scientists have created embryonic stem cells in mice without destroying embryos in the process

In August 2006, researchers at Kyoto University reported that by activating four genes in a mouse skin cell, they could reprogram that cell into a pluripotent state resembling an embryonic stem cell. However, the resulting cells were limited when compared with real embryonic stem cells, and the Kyoto team was unable to generate live mice from these cells. *6 June 2007*

⇨ The above information is reprinted with kind permission from the Massachusetts Institute of Technology. Visit http://web.mit.edu for more information.

© MIT

Stem cell milestones

Information from the Medical Research Council

Scientific

1957 First successful intravenous infusion of bone marrow in patients receiving radiation and chemotherapy, Mary Imogene Bassett Hospital, USA.

1978 First IVF baby born, following fertilisation of human eggs outside the body by scientists at Cambridge University, UK.

1981 Successful cultivation of mouse embryonic stem cells at the University of California, USA, and Cambridge University, UK.

1994 Corneal cells from donors successfully used to treat patients with burns to the cornea by researchers at Chang Gung Medical College in Taiwan.

1996 A sheep is cloned using cell nuclear replacement at the Roslin Institute, Edinburgh University, Scotland.

1998 First human embryonic stem cell lines are created at the Wisconsin Primate Research Center, USA.

1999 Neural stem cells multiplied and reintroduced into a Parkinson's patient cause 50 per cent improvement in motor tasks, Cedars-Sinai Medical Centre, USA.

1999 Scientists generate different types of blood cell from adult neural stem cells, Neurospheres Ltd, Canada and Consiglio Nationale Recherche, Italy.

2002 Research shows that adult stem cells can sometimes specialise into unrelated cell types, such as nerve and blood cells, University of Minnesota Medical School, USA.

2003 Researchers generate the UK's first human embryonic stem cell line, King's College London, UK.

2004 First stem cell lines deposited in UK Stem Cell Bank by scientists from King's College London and the Centre for Life in Newcastle.

2005 Scientists use stem cells to create the first mouse model of human Down syndrome, MRC National Institute of Medical Research and University College London, UK.

2006 Led by the MRC, the International Stem Cell Forum (ISCF) sponsors an international project coordinated by Sheffield University to characterise 65 stem cell lines.

UK regulatory

1982 The Warnock Committee is formed to examine the moral questions surrounding assisted reproduction and embryo research.

1984 The Warnock report backs human embryo research into reproductive related areas but advises tight regulation.

1990 The Human Fertilisation and Embryology (HFE) Act is passed by the Houses of Parliament.

1998 The HFE Authority and the Human Genetics Advisory Commission recommend that cell nuclear replacement (CNR) is investigated for therapeutic purposes but not for reproductive cloning.

2000 A report by the Donaldson Commission recommends that research using human embryos (created by IVF or by CNR) to increase understanding of human disease and disorders, and their cell-based treatments, should be permitted subject to the controls in the 1990 HFE Act. Parliament prohibits reproductive cloning.

2001

2002 A House of Lords Select Committee concludes that stem cells have great therapeutic potential and that research should be conducted on both adult and embryonic stem cells.

2002 A Steering Committee, chaired by Lord Naren Patel, is set up to oversee the development of a UK Stem Cell Bank and to establish codes of practice for the use of human stem cell lines.

2003 The UK Stem Cell Bank, funded by the MRC and the Biotechnology and Biological Sciences Research Council, is initiated at the National Institute for Biological Standards and Control.

2003 The International Stem Cell Forum (ISCF) – made up of 19 funders of stem cell research from around the world including the MRC – is established to encourage international collaboration and funding support for stem cell research.

2004 The ISCF begins a review of global ethics and regulation relating to stem cell research.

2005 The UK Government sets up the UK Stem Cell Initiative, with the aim of working with the public and private sectors to draw up a ten-year vision for UK stem cell research.

2006 The Human Tissue Act 2004 and Human Tissue (Scotland) Act 2006 come into force. The Human Tissue Authority regulates the storage of all cells for human application across the UK, as one of the competent authorities under the EU Tissue and Cells Directive.

2006 The Biotechnology and Biological Sciences Research Council leads the research councils and relevant government departments in establishing a UK National Stem Cell Network to encourage links between researchers and to encourage development of an integrated national stem cell research community.

2006 On behalf of the ISCF, the Australian National Health and Medical Research Council produces a report on intellectual property rights related to stem cell research across the world, which will be key in encouraging further research and development worldwide.

⇨ The above information is an extract from the MRC's document 'Stem cells – MRC research for lifelong health', and is reprinted with permission. For more information, please visit www.mrc.ac.uk

© Medical Research Council 2006

Heart tissue from stem cells

Technion researchers succeed in creating in the lab beating heart tissue from embryonic stem cells

The researchers also succeed in creating blood vessels inside the tissue, which will enable its acceptance by the heart muscle.

Technion researchers from the Faculty of Bio-Medical Engineering and the Rappaport Faculty of Medicine have succeeded in creating in the laboratory beating heart tissue from human embryonic stem cells. Moreover – they have succeeded in creating blood vessels in the tissue, which will enable its acceptance by the heart muscle.

The researchers are Dr. Shulamit Levenberg and Prof. Lior Gepstein. They were joined by doctoral student Oren Caspi and master student Ayelet Lesman.

The prestigious scientific journal Circulation Research reports in its on-line issue on two innovations in the researchers' work: one, the use of human embryonic stem cells, and two, the creation of a vascular system in the tissue, which is critical for its acceptance by the body.

'Without this system, acceptance could be prolonged and the cells could die during this time period,' explains Dr. Levenberg. 'In our work, we demonstrated the importance of the endothelial cells (cells that build blood vessels), which encourage differentiation of the heart cells and their organization, in addition to their multiplication. That is – it is important to create heart cell tissue, with all its component cells, in this case the endothelial cells, heart cells and cells that support the blood vessels.'

The Technion researchers created the heart tissue in the laboratory by differentiating human embryonic stem cells into heart muscle cells and endothelial cells and growing them together with embryonic supporting cells (fibroblasts). The growth was done in three dimensions on a porous, biodegradable scaffold that the Technion researchers also created in their laboratory. In the future, they will examine the possibility of implanting the tissue in a heart, in order to see if the blood vessels in the engineered tissue will improve acceptance of the new tissue and its connection to the vascular system.
14 January 2007

⇨ The above information is reprinted with kind permission from the Technion Israel Institute of Technology. Visit www.technion.ac.il for more information.

© Technion Israel Institute of Technology

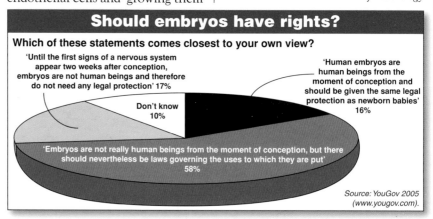

Should embryos have rights?

Which of these statements comes closest to your own view?

'Until the first signs of a nervous system appear two weeks after conception, embryos are not human beings and therefore do not need any legal protection' 17%

Don't know 10%

'Human embryos are human beings from the moment of conception and should be given the same legal protection as newborn babies' 16%

'Embryos are not really human beings from the moment of conception, but there should nevertheless be laws governing the uses to which they are put' 58%

Source: YouGov 2005 (www.yougov.com).

Women may be able to grow own sperm

By Roger Highfield,
Science Editor

Women could one day grow their own sperm, says a scientist who today claims to have turned bone marrow into early-stage sperm cells.

His team is now studying how to grow fully fledged sperm from bone marrow as a means to restore fertility in young men who have undergone cancer treatments. It could be five years before trials start.

Prof Karim Nayernia and his colleagues have completed similar experiments with female mice.

Within a few months, they expect to see if the method can be extended so that early-stage sperm can in principle be made from women, too.

Along with cloning, this could mark the second technique that makes men redundant.

The derivation of 'putative' early-stage sperm from four male volunteers is published today in the journal Reproduction: Gamete Biology by Prof Nayernia of the Northeast England Stem Cell Institute, Centre for Life, Newcastle upon Tyne.

Some experts urged caution about his claims, questioning the safety of using sperm derived from stem cells, and one was sceptical about the idea of women making sperm.

For the experiment, which was undertaken at Göttingen University, in Germany, Prof Nayernia's team took bone marrow from male volunteers and isolated a 'parent cell'. These stem cells have been found to grow into other tissues such as muscle.

In laboratory tests, they persuaded the stem cells to develop. The resulting cells look ordinary but genetic markers suggested that they were partly-developed sperm cells, otherwise known as 'spermatogonial' stem cells.

In most men, spermatogonial cells eventually develop into mature sperm but this progression was not achieved in this experiment, said Prof Nayernia.

He said his work was backed by a study of mice by Prof Ronald Swerdloff at the University of California, Los Angeles, who reported in the American Journal of Pathology that it was possible to turn these stem cells into early sperm cells.

Prof Nayernia said: 'We're very excited about this discovery. Our next goal is to see if we can get the spermatogonial stem cells to progress to mature sperm.'

He is to ask for permission to carry out similar experiments with human bone marrow in Newcastle, where the Centre for Life unveiled the world's first cloned human embryos.

As for growing sperm from females, 'we have evidence that it is possible, in the mouse at least'. However, more work was required to confirm this.

Prof Nayernia is concerned that the Government could outlaw treatments based on such work. A White Paper argued that the use of artificial gametes – eggs and sperm – would 'raise profound new possibilities such as the possible creation of a child by combining the genetic material of two women'.

Prof Robin Lovell Badge, of the National Institute of Medical Research in Mill Hill, London, said there were fundamental reasons why female bone marrow could not be converted into sperm.

Men have a Y and an X chromosome (bundles of genes in cells), whereas women have two X chromosomes. Prof Lovell Badge said that a Y was essential for sperm while having two X chromosomes was 'incompatible with making sperm'.

'The only true defining feature of a germ cell is its ability to undergo meiosis – the form of cell division that leads to sperm or eggs having just one of each chromosome rather than a pair, as found in all other cells.

The authors say they have not yet determined whether their cells can do this, but this is essential if they are to justify their claims.'

He said the paper contained 'several misleading statements', cited a 2006 paper that was 'questionable' and rested on the use of molecular markers to reveal the 'putative' sperm cells, which is a long way from the ultimate test of using sperm made this way to create an offspring.

Prof Harry Moore at the University of Sheffield said: 'We need to be very cautious. Unfortunately, these stem cell manipulations can lead to permanent genetic changes, which would make them unsafe to use.'

Josephine Quintavalle of Comment on Reproductive Ethics, CORE, said: 'How ironic that one of the few reproductive novelties the Government wants to ban is the very one that CORE could make a case for. But infertile men shouldn't get excited. There is far too much hype in this paper. As to growing sperm from women? As any A-level biology student would question, 'Where are they going to get the Y chromosome from?' '

13 April 2007
© Telegraph Group Limited, London 2007

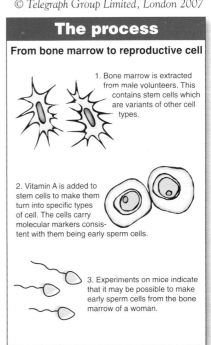

The process
From bone marrow to reproductive cell

1. Bone marrow is extracted from male volunteers. This contains stem cells which are variants of other cell types.

2. Vitamin A is added to stem cells to make them turn into specific types of cell. The cells carry molecular markers consistent with them being early sperm cells.

3. Experiments on mice indicate that it may be possible to make early sperm cells from the bone marrow of a woman.

Dolly

Information from the Roslin Institute

Birth

Dolly started her life, as with all other cloned animals, in a test tube. Once normal development was confirmed at 6 days, the embryo that was eventually to become Dolly was transferred into a surrogate mother. Pregnancy was confirmed by an ultrasound scan at about 45 days of gestation and the pregnancy was monitored closely for the remaining 100 days. The pregnancy went without a problem and Dolly was born on the 5th July 1996. Unlike many cloned animals who often have neonatal problems at birth, Dolly was a normal vigorous lamb and was standing and suckling unaided within minutes. The animal technicians were aware that this was an important lamb and critical to the research team that had produced her but they were completely unaware of the impact she would finally have.

Announcement to the world

The birth of Dolly was kept under wraps until the publication of the results could be prepared. Once these results were released, the full impact of the discovery became plain to all the animal carers as the world's press descended on Roslin. Most staff thought that this initial interest would be brief and quickly fade, but that was not the case and the press, in all shapes and forms, regularly visited Dolly for one reason or other for the rest of her life with interest peaking every time there was any concern over her health or when another species was cloned.

Was Dolly already 'old' at birth?

The first wave of increased press coverage occurred when, at one year old, tests revealed that Dolly's telomeres were shorter than those expected for sheep of that age.

Telomeres are sections of DNA found at the end of each chromosome. When the chromosomes are replicated during cell division a small portion of the telomeres are lost. They get shorter and shorter as more cell divisions occur and as animals age. This led to speculation in the press that animals cloned from cells obtained from an adult animal would age prematurely and die early. This was later shown to be untrue and in fact the telomere length is restored during the cloning process. Although Dolly's telomeres appeared shorter than other sheep of a similar age they certainly were not of an old animal. Extensive health screens carried out at the time failed to identify any abnormality with Dolly that would suggest premature aging.

Dolly's family

In an attempt to allow Dolly to have as normal as life as possible and to demonstrate that she was physiologically normal it was decided that she should be allowed to breed. A small welsh mountain ram was selected as her mate and between them they successfully produced 6 lambs. Their first, Bonny, was born in the spring of 1998. Twins followed the next year and triplets the year after that.

Dolly's arthritis

Press interest in Dolly had quietened down for a while until, in the autumn of 2001 and at the height of the Foot and Mouth outbreak in the UK, Dolly was seen to be walking stiffly. An x-ray was essential, and after getting permission from DEFRA for a special movement licence, Dolly was transported to the Royal (Dick) School of Veterinary Studies (R(D)SVS). Here a general anaesthetic was administered and thorough investigations of her lameness carried out. The x-rays confirmed that Dolly did indeed have arthritis. This was a blow to everyone and again fuelled the suspicion that cloned animals were destined to age prematurely. The cause of the arthritis was never established but daily anti-inflammatory treatment resolved the clinical signs within a few months.

Dolly's final illness

Although the arthritis was a concern for the animal carers at Roslin, a much

more serious problem was feared. In January 2000, another cloned sheep, Cedric, died. The post mortem revealed that Cedric had died of sheep pulmonary adenomatosis (SPA). This disease is caused by a virus that induces tumours to grow in the lungs of affected animals. The virus is infectious and spreads from sheep to sheep by close contact. The disease is incurable.

The animal carers were clearly concerned that Dolly might be infected with the same virus but hoped that she could still be free of infection as Cedric had been housed in a separate pen to Dolly. She was immediately placed in isolation; she could see and hear other sheep and was housed with her first lamb Bonny but was prevented from mixing with all other sheep. Even worse news soon followed when in March, Morag, one of the first cloned sheep from cultured cells, succumbed to the same disease. She had been housed with Dolly for many years so the chance that Dolly was not infected were slim. When, in September one of Dolly's second litter of lambs was also diagnosed with SPA, it became certain that Dolly herself must be infected. At that point it was decided that her isolation served no purpose and Dolly was returned to the flock of cloned sheep.

SPA was a difficult disease to cope with; there were no blood tests available to confirm the diagnosis and no effective vaccines or treatments. The most important task was to ensure that, if Dolly did develop the disease, she should not be allowed to suffer. In addition to her regular daily health checks by the animal care staff, veterinary examinations were increased and her weight was measured weekly. Dolly was as far as could be established perfectly well. She remained healthy until Monday the 10th February 2003 when an animal care worker reported that he had noted Dolly coughing. Full veterinary examinations and blood tests were conducted but failed to establish a diagnosis. Further investigations were necessary and with the kind co-operation of the Scottish College of Agriculture and the R(D)SVS a CT (computer tomography) scan was carried out

on Friday the 14th February 2003. The scan confirmed our worst fears, tumours were growing in Dolly's lung. Since a general anaesthetic had been necessary to perform the CT scan it was decided that it would be best if Dolly did not regain consciousness and an overdose of an anaesthetic agent was administered to end her life.

Why was Dolly created?

The development of the cloning technology was an extension of Roslin Institute's interest in the application of transgenic technology to farm animals. Transgenic mice have been available since the early 1980s. They are produced by a very sophisticated method of genetic modification through a technology using embryonic stem cells. Cells in culture can be genetically modified in very precise ways: removing genes, substituting one gene for another, introducing a single base pair change in the genetic code. In mice it was possible to genetically modify these cells, introduce them into a mouse embryo and the resulting mice that are born would be chimaeric (containing some normal cells and some genetically modified cells). At least some of the offspring of these chimeras would contain the very precise genetic modification. Since embryonic stem cells had not been isolated from farm animals, this method of genetic modification was not available. Cloning was therefore a potential alternative way of achieving the same end.

Why was Dolly important?

The birth of Dolly overturned the assumption among scientists that the whole process of differentiation was irreversible. We all start life as a single cell; the fertilised egg. The cell divides and multiplies and by the time we are born, there are maybe 200 different cell types, each containing the same DNA, the same 30,000 to 40,000 genes, but each has differentiated into a particular role. That role is determined by the proportion of active genes within the cell that determines whether the cell is for example a liver cell or a nerve cell. A presumption among cell biologists was that this was a

one-way process of progressive and permanent change. What Dolly demonstrated was that it is possible to take a differentiated cell and essentially turn its clock back – to reactivate all its silent genes and make the cell behave as though it was a recently fertilised egg.

Dolly was also important because she captured the public imagination. A clone – a copy – has been a very discernible strand within science fiction. The idea that there might be an exact copy of yourself somewhere around is a theme that has been pursued often and the prospect that it might be possible to clone a human being excited a lot of speculation and interest.

A sheep

What is the long term significance of Dolly?

At the moment that's difficult to say. The practical applications of cloning or copying livestock seem relatively limited. The likelihood is that the longer lasting benefit will be in the change in perception about biology. Our understanding now is that the cells in our bodies are a lot more plastic than we previously thought and it may be that as we understand more about repair processes for various organs and tissues, we might find that this understanding informs research that is able to augment the body's normal repair mechanisms. It may well prove to be an important factor in stem cell research and allow the derivation of stem cells from tissues other than early human embryos. This would alleviate the reservations that many people have about the use of human embryos for research or therapeutic purposes.

⇨ The above information is reprinted with kind permission from the Roslin Institute. Visit www.roslin.ac.uk for more information.

© Roslin Institute

Where Dolly went astray

Ten years ago the first mammal cloning seemed to herald a new era of medicine – then nothing happened. Robin McKie, who broke the story, meets the pioneer who says Britain let another breakthrough slip away

It was a breakthrough decades ahead of its time. On 23 February, 1997, the world learnt that British scientists Ian Wilmut and Keith Campbell had created the first clone of an adult mammal. 'They have taken a cell from a sheep's udder and turned it into a lamb,' ran The Observer's front-page story. Dolly the Sheep had arrived.

The creation of Dolly – at the Roslin Institute, the agricultural research centre near Edinburgh – opened up the prospect of an era of new medicines and treatments for conditions such as Alzheimer's and Parkinson's. It also triggered a fierce debate about the prospects of cloning humans and creating armies of Saddam Hussein doppelgangers.

Yet, despite the fuss, neither medicines nor cloned dictators have materialised. For a technology that promised to transform the world, this absence is startling. Indeed, most recent headlines have focused on South Korean scientist Hwang Woo-suk. He once claimed to have cloned human embryos and extracted stem cells, which have the potential to be used as life-saving medicines. However, Hwang has since been revealed to be a fraud.

Controversy, but not results: not much of an epitaph for Dolly. She died in 2003 and her stuffed remains are now displayed in Edinburgh's National Museum of Scotland. Will we soon wonder what was so special about her? Why have we not seen more scientific breakthroughs emerging – particularly in Britain?

It is an issue that vexes Wilmut. At present the best use of cloning has been the creation of a herd of cows that can make human antibodies, he believes. 'This has really important implications,' he said in an interview with The Observer last week. 'If a virus or a cancer gets past a person's immune system, he or she is in trouble. Soon we may be able to use cattle to provide antibodies that would pick up such a tumour or microbe. It has great potential.' The trouble is that the herd is the creation of Jim Robl of Hematech, a company based in the United States.

In fact, Wilmut has moved on from the Roslin Institute, where he did his pioneering work, to become director of Edinburgh University's centre for regenerative medicine. In effect he has turned his back on agricultural research and moved into medical science. A major motivation seems to have been the failure of Roslin to take advantage of his cloning work for use in agriculture, though Wilmut is reticent on this issue.

A few years ago the companies set up by the institute to exploit his cloning techniques were sold to the US biotechnology giant, the Geron Corporation. With those companies went the rights for Wilmut and Campbell's cloning technology. As a result, if a scientist or company wants to make a cloned animal as a part of a commercial enterprise today they have to pay Geron for the privilege. Even Wilmut would have to stump up – though only if he wanted to use it for commercial purposes. Scientists doing basic research do not have to pay.

'We are not entrepreneurial in this country,' Wilmut said. 'This is another technology that has walked, which means we miss out from the point of view of return on investment, on employment and in getting good access to profits.'

Thus the rights to use cloning techniques, and the best example of its exploitation, reside on the other side of the Atlantic. It is not a healthy position.

On the other hand, Britain retains the whip hand when it comes to the use of cloning in medicine for humans. Human embryos can be made by cloning and then be used as sources of stem cells, which have great promise as treatment for diseases such as Parkinson's and Alzheimer's. However, work like this is banned in federal laboratories in the US, giving Britain a clear advantage in ground-breaking research in this field.

A classic example of the potential cloning holds for humanity is provided by one of the crippling – ultimately fatal – illnesses, motor neurone disease. Wilmut was recently given permission to create human embryos from the DNA of victims of the disease, which strikes people at an average age of 56 and affects about 1,200 people in Britain every year. The aim of his research was simple. He intended to grow embryos taken from motor neurone patients, then remove the stem cells – the cells that eventually grow into the bodies' different tissues. In this way, key insights into the development of the disease would be gained, raising hopes of developing new drugs.

However, the gods of cloning again conspired against Wilmut. He was persuaded by Woo-suk Hwang – then hailed as the world's greatest cloner – that this approach was wrong and doomed to failure. So Wilmut gave up the programme and let his licence lapse. 'I should have known better,' he said. 'We'd visited Hwang in Korea on a couple of occasions and he had all these invalids in wheelchairs lined up in the audience in front of the platform that he was speaking from. He would tell them he would have them walking in no time. It really wasn't on.'

Now Hwang has been disgraced and Wilmut's approach vindicated. However, the licence has expired, which means his research plans into motor neurone disease have been set back by about five years. However, he remains determined. Indeed, his commitment is touching. In America, he pointed out, motor neurone disease is known as Lou Gehrig's disease, after the baseball star who died of it in 1941. In this country the disease should, by the same logic, be known as Jimmy Johnstone's disease. Johnstone was the Celtic winger who helped his team win the European Cup in 1967 and whose dazzling bursts on the right earned him the Glaswegian accolade of 'Wee Jinky'. Last year Johnstone died of motor neurone disease.

'I had the privilege of meeting him and his wife a few months before he died. He and I were roughly the same age, but he was completely paralysed from the neck down.' The recollection stopped Wilmut in his tracks, and for a couple of minutes he seemed on the cusp of tears. 'I am not normally lost for words. It was extremely moving meeting someone like that. Why are we not doing what we can for them?'

As head of his university's new regenerative medicine centre, Wilmut intends to do just that and aims to regalvanise his efforts in seeking breakthroughs in the treatment of motor neurone disease: 'That would be the best possible way to celebrate Dolly's life.'

Cloning will ultimately be extremely valuable to medicine, Wilmut added. It is just proving hard to exploit. It took 270 attempts to create the embryo that resulted in Dolly. 'That represents an efficiency of 0.3 per cent. Today we have got that up to about 3 per cent. But it is still very low.'

There are other reasons for our failure to follow on swiftly from Dolly's creation, however, as Simon Best of the UK Bioindustry Association pointed out: 'Dolly's creation was simply so far ahead of its time it caught scientists out. If you had taken a poll of scientists in 1997, the vast majority would have said there was no prospect of cloning a mammal for another 15 to 20 years.

'Wilmut and Campbell caught everyone by surprise. This field was considered so uncompromising only a handful of scientists were working in it. There were therefore no promising young scientists, expert in the field, to take up this work and follow through with new breakthroughs. That is why we are still having to wait.'

Look at other great scientific discoveries and you see a similar picture, added Best. When Crick and Watson discovered the structure of DNA, it took more than 20 years for their work to be translated into biotechnology products including human growth hormone, blood clotting treatments and other drugs.

And the science behind DNA's structure was much less revolutionary than that involved in Dolly's creation. Crick and Watson indulged in a great deal of scientific detective work to make their discovery, of course. But Wilmut and Campbell carried out hundreds of highly complex biological experiments before they succeeded. To create Dolly they removed a cell from an adult sheep. Then they took an egg from another sheep, scooped out its DNA and inserted the first sheep's nucleus. After a series of chemical treatments, they then placed this artificial embryo into a third sheep and coaxed it to grow into Dolly. This is cloning or nuclear transfer technology.

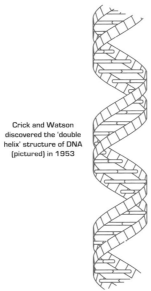

Crick and Watson discovered the 'double helix' structure of DNA (pictured) in 1953

'Until then, we all thought it was impossible to reprogramme an adult cell and persuade it to act like an embryo,' said Cambridge cloning expert Roger Pedersen. 'Wilmut and Campbell showed that was wrong. They turned back the arrow of time. And we are still trying to come to grips with the implications.'

Wilmut looks an unlikely scientific revolutionary. He has an amiable, avuncular demeanour and describes himself as 'a rather shy, reserved middle-aged Englishman'. It is a fair description. His casual, bearded appearance makes him look more like a Yorkshire farmer than a researcher whose work frightened the living daylights out of politicians round the world.

'I hadn't expected all that fuss,' he admitted. 'Even in the middle of it, I thought it would all die down. I remember being on holiday, walking along a beach in Scotland, six months after the story broke, thinking: 'Oh, well, at least it is now all over and I can got on with my work.' But of course, it wasn't all over. It has never gone away. It just shows what I knew about my work's implications.'

Stem cell timeline

⇨ 1981 – Gail Martin at the University of California, San Francisco, and Martin Evans, University of Cambridge, isolate embryonic stem cells in mice for the first time.

⇨ 23 Feb, 1997 – Dolly the Sheep (born at the Roslin Institute) is revealed to the world. It is the first clone of an adult mammal.

⇨ July 1998 – Noto and Kaga, the first cloned cows, are produced by the Ishikawa Prefectural Livestock Research Centre, Japan.

⇨ 2001 – The UK becomes the first country to legalise therapeutic human cloning for research.

⇨ Feb 2003 – Dolly the Sheep dies.

⇨ April 2003 – The sequencing of the human genome is completed.

⇨ Feb 2004 – South Korean scientist Hwang Woo-suk and his team claim to have created 30 cloned human embryos and extracted stem cells.

⇨ August 2005 - The first dog clone, an Afghan hound called Snuppy, is born in South Korea, also created by Dr Hwang.

⇨ Jan 2006 – Dr Hwang's work exposed as fake.

⇨ July 2006 – President Bush uses his veto in order to kill a bill to increase federal funding of stem cell research.

18 February 2007

⇨ This article first appeared in *The Observer*

© *Guardian Newspapers Limited 2007*

Man or mouse

Information from Animal Aid

Hundreds of millions of genetically modified (GM) animals are used in medical research and in the testing of drugs and chemicals every year. In response to their increasing utilisation, Animal Aid has commissioned this report, which examines the GM animal phenomenon, and shows that they have made no tangible contributions to human health and medicine – and, for sound scientific reasons – are never likely to do so.

In this report we show that:

⇨ The annual total number of animal experiments in the UK was in steep decline from 1975 to 1985, during which time it fell by around 40%. The use of 'normal' (non-GM) animals has continued to decline at a similar rate since 1985, as scientists have abandoned this approach due to its lack of success and relevance to human health. The increased use of GM animals has, however, reversed the overall trend. Because of GM animals, the number of animal experiments is once again increasing, with GM animals constituting almost one third of all procedures.

⇨ GM research has failed to live up to its promise. Rather than 'fixing' the old non-GM animal models by genetically manipulating them to resemble more closely human beings and human diseases, this process has revealed itself to be hopelessly inefficient and crude. Some 70% of the time when a GM animal intended to replicate a human disease is created, it does not 'perform' as expected. Put simply the actions of, and interactions between, our genes are much more complex than first thought: minor genetic differences between individual humans combine to be greater than the sum of the parts. To expect to derive useful information from another species altogether is extraordinary.

⇨ Modelling human diseases with GM animals has been disastrous. Cystic fibrosis (CF) and Alzheimer's disease (AD) have been extensively researched using GM animals, which has served only to confound our knowledge of them and to impede progress. CF affects the pancreas in almost all human sufferers, and kills via lung infections. Mice with 'identical' genetic mutations to human CF patients do not show these effects, but rather die early from intestinal blockages not seen in humans. GM animals have failed to replicate the pathology of human AD: GM mice with identical brain pathology to human AD sufferers show no or only slight effects, and have failed to shed any light on the function of genes strongly linked with human AD. Other GM animal models of human diseases such as Parkinson's and diabetes have failed similarly. Nor should we expect a different outcome from currently 'fashionable' projects, such as the new genetically manipulated mouse 'model' of Down's Syndrome. In contrast, human-specific research methods continue to make significant contributions to medical progress in these areas.

⇨ GM animals have failed in all fields in which they have been applied. Animals have been engineered to be more predictive models in toxicology testing, but this has not been realised. Engineering animals to grow organs for human transplants has been futile and is considered highly dangerous because of the risk of 'novel' disease organisms passing from 'donor' animal to human recipient and from there into the wider population. While some animals have been created who can produce drugs in their milk, for example, there are questions surrounding the need for this. There are also serious concerns regarding the welfare of the animals involved. Cloning of various animals has been reported in the media, but continues to be extremely inefficient, with poor survival rates, significant welfare problems and a high incidence of defects and abnormalities.

⇨ Human-specific research, utilising tried and tested methods that have contributed greatly to medical progress, along with cutting-edge technologies, are the only way to achieve safe, efficient cures and treatments for human diseases in the shortest possible time. Increasing numbers of scientists and doctors agree, and are turning to research methods involving – for example: human tissue and cells, computer modelling, DNA 'microarray' chips, microfluidics and advanced scanning technologies – shunning animal-based approaches.

⇨ On balance, GM animals have made a negative contribution to human medicine. At the same time, it must be recognised that competition for medical sector funding remains acute. An important choice confronts society: the choice between more resources directed at animal-based research with a fruitless track record – or support for work that is directly relevant to the patients of today and tomorrow. Animal Aid believes that a comprehensive scientific evaluation of animal-based approaches to research is absolutely imperative to achieve real medical progress.

⇨ This information is the executive summary of the Animal Aid report 'Man or Mouse' and is reprinted with permission. For more information, please visit the Animal Aid website at www.animalaid.org.uk

© Animal Aid

Clones and factory farming

Cloning opens door to 'farmyard freaks'

Moves to clone and genetically modify farm livestock have opened the door to the creation of 'Farmyard Freaks', experts have warned.

News that the daughter of a US clone cow has been born on a British farm has moved the issue from science fiction to consumer reality.

A former government adviser has painted a nightmarish picture of 'zombie' and fast-growing supersize animals.

Professor Ben Mepham, of Nottingham University, said the impact of bio-engineering, creating GM and cloned animals, is huge.

Factory farming techniques, most commonly used with pigs and chicken, often involve keeping animals confined in cramped conditions.

For pigs, who are highly intelligent, these conditions can lead to stress and aggression.

However, GM scientists are actively investigating ways to remove the stress and aggression gene from animals, effectively turning them into complacent zombies.

The professor said it might become technically possible to produce 'animal vegetables' – beasts which are 'highly prolific and oblivious to their physical and mental status'.

However, he argued that while this could reduce the pain and stress of factory farming, this did not mean it should be allowed to develop without question.

The professor of applied bioethics warned that many of the GM experiments on animals have resulted in cruelty, producing mutants or animals which grow so large in the womb that they can only be surgically removed.

He said: 'The question of whether humanity should take it upon ourselves to alter animals by GM, involving in many cases mixing the genes of different species – and sometimes those of human origin – is undoubtedly critical for many people.'

The professor said that religious groups would see it as 'an attempt to usurp God's role' while others would be unhappy about 'so fundamentally altering the natural order'.

Prof Mepham, is a former member of the Government's Agriculture, Environment Biotechnology Commission (AEBC).

In 2002, the Commission called on the government to set up a regulatory body to police developments such as GM and clone farming.

However, this was ignored by ministers, who subsequently scrapped the AEBC after it issued a number of reports challenging government policy in areas such as GM crops and food.

The AEBC called for a ban on the creation of 'intrinsically objectionable' creatures – such as pigs and cows modified not to feel stress in factory farming conditions. And it demanded separate farming and labelling of food from these creatures to allow consumers to make a choice about what they are eating.

In 2002, the AEBC said the need to have in place a regulatory regime in place was 'urgent' in order to prevent a repeat of the GM crop debacle.

In that case GM plants were already in British shops before there had been sufficient research about the impact on human health or the environment.

Despite these clear warnings, the government's food and farming department, DEFRA, refused to set up any kind of watchdog.

The result is that meat and milk from GM or cloned animals could be arriving on dinner plates in as little as two years.

The executive director of the Food Ethics Council, Dr Tom MacMillan, said: 'Cloning raise animal welfare concerns, both for the clones and for their parents.

'It also underlines how far removed industrial food production is from what consumers actually want.'

Daily Mail, 11 January 2007
© *Associated Newspapers Ltd 2007*

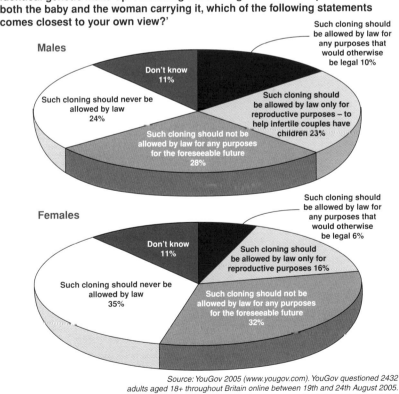

Should we clone humans?

Respondents were asked 'Cloning is the creation of a human, an animal or a plant from a single "parent". The clone is a "carbon copy" of its parent, with identical genetic make-up. Assuming the cloning of babies was proved safe for both the baby and the woman carrying it, which of the following statements comes closest to your own view?'

Males

- Such cloning should be allowed by law for any purposes that would otherwise be legal 10%
- Such cloning should be allowed by law only for reproductive purposes – to help infertile couples have children 23%
- Such cloning should not be allowed by law for any purposes for the foreseeable future 28%
- Such cloning should never be allowed by law 24%
- Don't know 11%

Females

- Such cloning should be allowed by law for any purposes that would otherwise be legal 6%
- Such cloning should be allowed by law only for reproductive purposes 16%
- Such cloning should not be allowed by law for any purposes for the foreseeable future 32%
- Such cloning should never be allowed by law 35%
- Don't know 11%

Source: YouGov 2005 (www.yougov.com). YouGov questioned 2432 adults aged 18+ throughout Britain online between 19th and 24th August 2005.

Milk from cloned cows leaks into UK

Dundee Paradise is evidence that clone farming – designed to deliver supersize cows producing an astonishing 70 pints of milk a day – has arrived in Britain.

Her birth last month exposed glaring gaps in the Government's system for policing livestock farming.

It raises the prospect of milk and meat from the offspring of clones reaching the shops without proper safety checks.

Though not a clone herself, Dundee Paradise is the daughter of a clone. Her mother was created in the U.S. using cells from the ear of a champion dairy Holstein.

Dundee Paradise herself began life in an IVF lab. She was flown to the UK in a batch of five frozen embryos, implanted in a surrogate mother and successfully delivered at a Midlands farm on December 2.

Both the food and farming ministry Defra and the Food Standards Agency admitted last night that they had known nothing about the calf's birth, or even the arrival of the batch of embryos last year.

Three years ago Defra rejected advice from its own experts to set up a safety assessment and policing regime. It means there are no laws preventing farmers from rearing clones or their offspring.

Supporters of cloning believe Dundee Paradise and animals like her could provide the nucleus of Britain's future dairy herd. The calf was valued at £14,000 even before she was born.

But the use of clones, sometimes dubbed Frankenstein Farming, is extremely controversial.

Critics say there are serious welfare issues. Clones and their offspring are known to be at risk of dying young – one of the batch of embryos miscarried at seven months – and having malformed organs.

There are also ethical questions about meddling with the building blocks of life simply to produce more cheap food. Perhaps the most alarming element is that the government has allowed clone farming to arrive in this country without proper public consultation or debate.

Compassion in World Farming, which has opposed the cloning and genetic modification of farm animals, said the public would be 'horrified'.

Policy adviser Peter Stevenson said: 'People will be astonished and appalled that clone farming is not a nightmare for the future, it already exists on a farm in this country.

'The history of cloning is littered with examples of animals that are unhealthy or die quite early. The whole process involves invasive procedures that cause suffering.'

He pointed to Dolly the sheep, the first mammal clone. Born in a British farm laboratory in 1997, she died young in 2003 from a lung disease. She was already suffering painful premature arthritis.

Mr Stevenson said: 'We would urge the farming industry to call an immediate halt to this and we would urge consumers to refuse to buy meat and milk from these sources.'

He said it is not too late for the government to block the controversial technology.

Lord Melchett, policy chief of the Soil Association, said: 'I cannot think of anything more likely to destroy the public's confidence in British food.

'High-yielding Holstein cows are already one of the biggest welfare concerns in farming because of the huge strain of producing vast quantities of milk. Government figures show that a third of dairy cows are killed after just one lactation because their bodies cannot cope with any more.

'Cloning would push this whole catastrophe one step further.'

Lord Melchett criticised the lack of a monitoring regime, saying: 'All the evidence suggests the government will refuse to take action unless it is provoked by some kind of crisis for human health.'

The farming industry believes the cloning of prize livestock will allow the creation of herds of supersize animals and boundless cheap food.

The giant Holstein cows can produce 30-40 per cent more milk than conventionally-bred animals. As a result, top animals can change hands for more than £50,000.

Dundee Paradise was born at a large farm in the Midlands.

Her mother is called Vandyk K Integ Paradise 2 and lives on a farm in Wisconsin.

She is one of three clones created by the U.S. biotech firm Cyagra Clone from cells taken from the ear of the multi award-winning cow Vandyk K Integ Paradise.

Embryos from such clone cows are highly sought-after in the U.S. and it appears the trade has now extended to the UK.

The company which owns Dundee Paradise is run by a father and son team who also have a major concrete products business.

They bought five embryos from the U.S. to be implanted in their Holsteins. The others are expected to be born in the next few weeks.

The company has refused to comment.

Could we be drinking milk from cloned animals?

An option to buy Dundee Paradise was sold to a member of the Bahrain royal family – Sheika Noora Bint Isa Alkhalifa – at a cattle auction held by Harrison & Hetherington in Carlisle in December.

The sheika, who owns prize pedigree Holstein herds in Devon, Wales and Bahrain, paid £14,000 to have first choice of the five clone-origin calves. This was before any of them were born.

The American owner of the clone mother cow confirmed the sale of embryos to the UK.

Mark Rueth, who farms in Oxford, Wisconsin, said: 'We sold some embryos to (the Midlands farmers) and they got a very good price for a calf. I can see the UK being a big market.

'The benefit of cloning is that when a donor animal is gone it means you are able to maintain and build a superior genetic base for the herd.'

Mr Rueth said the price of cloning works out at around £9,800 per calf. He expects this figure to come down.

Supporters of cloning argue that it could actually increase the lifespan of the animals involved. Bigger cows, will be better able to cope with the huge demands on their bodies.

The clones themselves are essentially used as breeding stock – embryo factories capable of churning out large numbers of high-value eggs and calves. For farmers who own a prize cow, the potential income is extremely tempting.

American cloning companies are busy making multiple copies of just about every top pedigree cow and bull in the land. Scientists are also working on pigs and chicken to produce model breeding stock.

The speed of the spread of clone farming is outpacing the regulatory system in Britain and the rest of the EU.

The U.S. Food and Drug Administration recently announced that it plans to allow the consumption of meat and milk from cloned animals without any need for labelling.

The FDA has angered health and consumer groups by arguing that these animals are effectively the same as those born naturally.

In 2004, Defra rejected advice from its own experts to set up a committee with legal powers to monitor and police attempts to introduce clone farming or genetically-modified animals.

The Department said such controls would stifle British science and innovation and impose unnecessary red tape and costs on the farming industry.

Now, just as the experts warned, clone farming has arrived in a way which cannot be controlled or monitored.

Defra insisted last night it was following EU rules which 'do not require us to differentiate cloned from normal embryos'. It said there was no plan for tougher UK-only rules.

The Food Standards Agency has been told by ministers to police the introduction of meat and milk from cloned animals and their offspring.

But it also knew nothing about Dundee Paradise.

⇨ This article first appeared in the *Daily Mail*, 9 January 2007

© 2007 Associated Newspapers Ltd

Can we be sure no cloned animals are in food chain?

⇨ *Newspaper reveals existence in UK of calf born of cloned cow in the US*
⇨ *Defra accused of failing to regulate importation of 'Frankenstein' livestock*
⇨ *Cloned animals to date have suffered health problems*

Key quote
'From a scientific point of view, we would not have any additional concerns for the animals welfare over and above those we have for any cow. Any offspring from a cloned animal is considered to be the same as any other naturally born cow.' – DEFRA spokesperson

Standing at just waist height to her proud owner, six-week-old Dundee Paradise looks as unremarkable as any other calf. But behind the distinctive flash that runs down the middle of her

By Alison Hardie

face, the black and white Holstein represents what many are describing as a 'horrifying' development that could destroy confidence in food and farming in Britain.

Dundee Paradise is the daughter of a cloned cow produced in the United States. The calf was born after five embryos were brought into the UK without the knowledge of the government.

Farmers want to use the cloning technology to develop supersize cows able to produce 70 pints of milk a day.

But are British consumers ready to drink the milk and eat the meat of animals which are the offspring of clones?

Questions are now being asked as to why the existence of the calf only came to light in the pages of a newspaper and why regulations on importing embryos are not tighter.

Campaigners last night said the Department of Environment, Food and Rural Affairs (Defra) has acted irresponsibly by ignoring recommendations by an independent body three years ago to regulate the importation of 'Frankenstein' livestock.

'People will be astonished and appalled that clone farming is not a nightmare for the future, it is absolutely horrifying to learn it already exists on a farm in this country'

Defra said it was investigating whether proper procedure was followed when the cow's embryo was flown into the UK.

Lord Melchett, policy director of the Soil Association, said the government's failure to impose regulatory controls on such a practice was 'inexcusable'.

British consumers had no desire to eat meat or drink milk from animals created in such a way, he said.

Peter Stevenson, policy advisor for Compassion in World Farming (CiWF), said: 'People will be astonished and appalled that clone farming is not a nightmare for the future, it is absolutely horrifying to learn it already exists on a farm in this country.

'It is doubly worrying that there is no safeguard in place to avoid serious animal welfare and ethical problems from the introduction of this 'Frankenstein' technology.

'The history of cloning is littered with examples of animals that are unhealthy or die quite early and the whole process involves invasive procedures that cause suffering.

'The UK government must act responsibly by stopping further introductions and setting up an independent ethical watchdog to oversee this area.'

Dundee Paradise is owned by a farmer in the Midlands, however at least three other siblings are expected to be born at undisclosed UK locations in the next few weeks.

Defra has admitted 'no specific EU regulations govern the import of cloned animals or embryos other than those health and welfare conditions that must be met for all embryos or animals'.

A spokesman said: 'Defra does not believe that any animal health and welfare regulations have been contravened in this case.

'This case refers to a cloned animal, not a genetically modified one, therefore the genetic material of the cloned animal would be an identical genetic replica of the original cow.

'From a scientific point of view, we would not have any additional concerns for the animals welfare over and above those we have for any cow. Any offspring from a cloned animal is considered to be the same as any other naturally born cow.'

The Food Standards Agency (FSA), which monitors and patrols the quality and providence of foods produced across the UK, admitted it had known nothing about the birth of Dundee Paradise.

But amid growing concern that meat and milk from cloned animals and their offspring was already in the food chain, the agency moved to reassure shoppers.

In a statement it said that products from cloned animals and their offspring fell under EU Novel Food regulations. And it added that anyone attempting to sell meat or milk from cloned animals or their offspring would first have to apply to the FSA for authorisation .

'No applications have been received to date for products derived from cloned animals,' a spokeswoman said.

In 2004 Defra rejected advice from its own experts to set up a committee with legal powers to monitor and police attempts to introduce in the UK cloned or genetically modified animals.

Without the proper safeguards in place, campaigners said last night, there is a very real potential that meat and milk from cloned animals or their offspring could be passed off as products from conventional livestock.

Concern about food produced from clones is also rife in the USA where hundreds of cloned cows, pigs and sheep have already been produced to provide the genetic fingerprint for the 'super herds and flocks' of the future.

In December last year the American Food and Drug Administration approved the sale of meat and milk from cloned animals.

The move was condemned by the Food Commission, Britain's independent food watchdog, as a 'giant step in the wrong direction'.

However, the director of the Roslin Institute, near Edinburgh, which produced Dolly the sheep, the first mammal clone, said consumers had nothing to fear from food from cloned animals or their offspring.

Dr Harry Griffin said: 'The calf in question is the offspring of a cloned cow which would itself have been cloned from a high-performing dairy cow which would have been one of the very best available in the US.

While Dolly was celebrated, her curtailed life was the first warning that cloned animals would often be in poor health

---

'However, the calf is certainly not a cloned animal itself because it went through the natural reproductive cycle. Offspring of clones produced this way would be expected to be absolutely normal.

'Until there is evidence that cloned animals pose a risk to humans then there is no need to change the status quo.'

Dolly's death sounded first warning of how genetic clock hits health

The early deaths of some of the world's first cloned animals have led to concerns about their suitability as part of the food chain.

Dolly the sheep became a superstar. Yet she was doomed to live a short life, dogged by arthritis and lung disease.

While Dolly was celebrated – her remains are now preserved and on permanent show at the Royal Museum in Edinburgh – her curtailed life was the first warning that cloned animals would often be in poor health.

Dolly was put down in 2003 at only six years of age; sheep of her breed are usually expected to live until 12.

Her 'cousin' Morag, a Welsh mountain ewe cloned in 1995, died of a respiratory infection.

When scientists announced in 2005 that they had produced the world's first cloned dog, called Snuppy, at first more attention was paid to his cute looks than the price that had been paid to produce him.

Later, the scientists behind his creation at Seoul University's College of Veterinary Medicine said they had obtained just three pregnancies from more than 1,000 embryo transfers into 123 recipients.

Of these, one miscarried. Two clones were eventually born, but one died of pneumonia.

One of the most serious problems plaguing cloning is the 'biological clock' which seems to be imbedded in the nuclei of cells, which can switch various processes on or off, and tells the cell itself to die. For example, it will enable processes to accelerate growth in young children, or produce sexual hormones at puberty.

In the case of Dolly, whose genetic

On our plates – does it matter if meat and milk from clones enters the food chain?

material was taken from a ewe, the clock effect was later linked to her health problems.

Scientists want to perfect techniques of animal cloning in order to have a reliable supply of special proteins, enzymes and hormones which can be used in the treatment of humans.

And they say that one day therapeutic cloning of human cells to cure illnesses could become as routine as IVF treatment.

11 January 2007

© *The Scotsman*

Cloned meat and milk 'safe'

Information from Farmers Weekly Interactive

A pilot study has shown that milk and meat from cloned cattle appear safe for human consumption, the BBC reports.

Scientists in the US and Japan found that meat and dairy products from a cloned bull and cow met industry standards.

The team said the results indicated cloning techniques were safe and could be used to boost food production, particularly in developing countries.

The study appeared in the Proceedings of the National Academy of Sciences.

Two beef and four dairy clones were used in the study, all derived from a single cow and a single bull.

The scientists, led by Jerry Yang from the University of Connecticut, compared the produce with that from comparable, conventionally bred animals.

They concluded that the study showed the produce to be within the range approved for human consumption, but stressed the research was at an early stage.

Critics have told the BBC, however, that the study is too small for firm conclusions to be drawn.

⇨ This article is reproduced with kind permission of Farmers Weekly Interactive: www.fwi.co.uk

© *Reed Business Information*

KEY FACTS

⇨ Cloning occurs in nature and can occur in organisms that reproduce sexually as well as those that reproduce asexually. In sexual reproduction, clones are created when a fertilized egg splits to produce identical (monozygous) twins with identical DNA. (page 1)

⇨ There are a variety of national laws on cloning; many other bills have been submitted and are currently under consideration. As of now, approximately 35 nations have adopted laws forbidding reproductive cloning. (page 3)

⇨ No one knows the psychological effects of discovering one was the twin of one of one's 'parents' or sibling. Am I just a copy of someone else who's already existed and not really 'me'? What would be my relationship to them? Since we have no sure way of knowing in advance, we surely do not have the right knowingly to inflict that risk on another person. (page 6)

⇨ At the early stage where research is focused, an embryo has few of the characteristics we associate with a person. It is a fertilised human egg, with the capacity to develop into a person, but its cells have not yet begun to form into specialist cells that would form particular parts of the body. (page 7)

⇨ 68% of people surveyed by YouGov felt that it was acceptable for 'spare' early embryos left over from fertility treatments to be used for medical research. (page 8)

⇨ Don't say cloning, say somatic cell nuclear transfer. That at least is the view of biologists who want the term to be used instead of 'therapeutic cloning' to describe the technique that produces cloned embryos from which stem cells can then be isolated. This, they argue, will help to distinguish it from attempts to clone a human being. (page 9)

⇨ Human eggs are in short supply – being also in demand for fertility treatment – and are expensive, costing about £3,000 per woman. Moreover, the procedure carries risks. Rabbit or cow eggs would enable scientists to do basic stem cell research more cheaply and safely. (page 9)

⇨ In biotechnology it is now possible to combine elements between organisms of different species. It is also possible to create cloned animals using parts of eggs from one species and nuclear genetic material from another. (page 10)

⇨ It may be said that any form of mixing violates natural boundaries – it breaks the species barrier. (page 12)

⇨ Hundreds of thousands of patients with diseases of the nervous system will miss out on potentially life-saving new treatments if regulators ban experiments using part-human, part-animal embryos, scientists have said. (page 14)

⇨ The aim of therapeutic cloning is to create embryonic stem cells for use in medical research and development of stem cell therapies. In contrast, the aim of reproductive cloning is to make a fully developed replica of the organism being cloned. Human reproductive cloning is illegal in the UK and the practice is considered to be unethical. (page 15)

⇨ 41% of people in Europe agreed that it is wrong to use human embryos in medical research. 41% of people disagreed. (page 19)

⇨ A positive 59% of Britons agree with controversial stem cell research, according to research by ICM conducted on behalf of the Guardian. (page 20)

⇨ Scientists believe stem cells could prove to be the ultimate body repair kit, with no need for donated organs, man-made joints or drugs to keep failing body parts working. (page 21)

⇨ 48% of people surveyed by YouGov felt that embryos should only be used in medical research for life-threatening diseases, such as cancer or heart disease, whether in adults or children. (page 24)

⇨ Scientists have created embryonic stem cells in mice without destroying embryos in the process, potentially removing the major controversy over work in this field. (page 26)

⇨ Women could one day grow their own sperm, says a scientist who claims to have turned bone marrow into early-stage sperm cells. (page 29)

⇨ On 23 February, 1997, the world learnt that British scientists Ian Wilmut and Keith Campbell had created the first clone of an adult mammal. Dolly the sheep had arrived. (page 32)

⇨ When Crick and Watson discovered the structure of DNA, it took more than 20 years for their work to be translated into biotechnology products including human growth hormone, blood clotting treatments and other drugs. (page 33)

⇨ 24% of males surveyed by YouGov felt that human cloning should never be allowed by law, compared to 35% of females. (page 35)

⇨ A pilot study has shown that milk and meat from cloned cattle appear safe for human consumption. (page 39)

GLOSSARY

Clone

The term clone derives from the Greek for 'twig', referring to the fact that a cutting from a plant will, when planted, grow into an identical copy of the parent plant. A clone, therefore, refers to one member of a group of identical entities – particularly, in recent years, the term has been used to refer to an organism that is a genetic copy of an existing organism. The term is applied by scientists not only to entire organisms but also to molecules (such as DNA) and cells.

Cord blood stem cells

Cells from the embryo present in blood vessels in the umbilical cord.

'De-differentiated' stem cells

Adult cells modified to become like stem cells. Not currently available, but being researched.

DNA

DNA stand for deoxyribonucleic acid. All human cells contain 23 pairs of chromosomes, one set inherited from each parent. Each pair of chromosomes are made up of a long strand of DNA, containing individual genes. This genetic material provides a blueprint for the cell itself and for the body in which it lives.

Dolly

Dolly the sheep represented a scientific breakthrough – she was the first animal to be cloned from an adult mammal, and was a celebrity in her lifetime as she overturned many previous assumptions about cloning among scientists, and captured the imagination of the public in general. She died prematurely from a disease of the lungs, and her stuffed remains are on display at the National Museum of Scotland in Edinburgh, the same city where she was originally cloned.

Embryo

An animal or plant in its earliest stage of development – in humans, this is the first eight weeks following conception. After this, it is referred to as a fetus. In the UK, the law states that spare embryos may only be used for medical research in the first 14 days after conception, after which they must be discarded.

Embryo cloning

In embryo cloning, individual cells are taken from an early embryo and each grown into a separate, identical embryo – it is a form of artificial twinning. The technique is used to make lots of identical copies of original embryos which have been genetically modified to produce human proteins. The adult animals which result provide life-saving treatments for thousands of people.

Hybrid

A cross between two species. In terms of cloning, the word 'hybrid' is applied to an embryo which is part human, part animal. This may also be called a chimera, or 'cybrid'. The process of creating a hybrid embryo involves taking an animal egg and emptying it of its nuclear DNA, before fusing it with DNA from a human cell. The resulting embryo would be 99.9% human. The purpose of this process is not to create an adult animal-human hybrid, but to produce embryos for stem cell research.

Somatic Cell Nuclear Transfer

Sometimes referred to as 'Dolly type' cloning, as this was the method used to create Dolly the sheep. When it is used to produce a new adult animal (as in Dolly's case), it may also be called reproductive cloning. This process begins with an adult cell (that is, one taken from an organism which has passed the embryonic stage of development). 'Somatic' denotes a body cell possessing a full set of chromosomes (and not half as in an egg or sperm cell). The nucleus of this cell is transferred to an egg which has had its nucleus removed and transferred into the uterus of a female host, where if it is implanted it can lead to pregnancy and eventually the birth of an animal which is the clone of the animal from which the somatic cell was originally taken.

Stem cells

There are two sources of stem cells: adult and embryonic. Whereas scientists think that adult stem cells are restricted to maintaining the health of the tissue where they are found, embryonic stem cells have the potential to turn into any cell type. If we can harness their adaptability, they might be a source of healthy tissue to replace that which is diseased or damaged in adults. However, because the embryo is destroyed during the process, the extraction and use of embryonic stem cells is highly controversial. Pro-life groups in particular have been campaigning against stem cell research, as they believe human life begins at conception and embryos should be afforded the same rights as adult humans.

Therapeutic or biomedical cloning

Also called non-reproductive cloning or research cloning, this involves the use of the Somatic Nuclear Cell Transfer technique without the intention of producing a new adult animal or plant. The hope is to produce new tissues or organs for people who are seriously ill with problems ranging from diabetes and Parkinson's disease to heart attacks and spinal injuries.

Therapeutically cloned embryonic stem cells

Embryonic stem cells generated following single cell nuclear transplantation. Cells generated will be genetically identical to the donor.

INDEX

Additional Resources

Other Issues *titles*

If you are interested in researching further some of the issues raised in *The Cloning Debate*, you may like to read the following titles in the **Issues** series:

⇨ Vol. 138 *A Genetically Modified Future* (ISBN 978 1 86168 390 8)

⇨ Vol. 134 *Customers and Consumerism* (ISBN 978 1 86168 386 1)

⇨ Vol. 98 *The Globalisation Issue* (ISBN 978 1 86168 312 0)

⇨ Vol. 120 *The Human Rights Issue* (ISBN 978 1 86168 353 3)

For more information about these titles, visit our website at www.independence.co.uk/publicationslist

Useful organisations

You may find the websites of the following organisations useful for further research:

⇨ CORE: www.corethics.org

⇨ GeneWatch UK: genewatch.org

⇨ Greenpeace: www.greenpeace.org.uk

⇨ Medical Research Council: www.mrc.ac.uk

⇨ Wellcome Trust: www.wellcome.ac.uk

ACKNOWLEDGEMENTS

The publisher is grateful for permission to reproduce the following material.

While every care has been taken to trace and acknowledge copyright, the publisher tenders its apology for any accidental infringement or where copyright has proved untraceable. The publisher would be pleased to come to a suitable arrangement in any such case with the rightful owner.

Chapter One: Human Cloning

Questions and answers on human cloning, © World Health Organization, What is cloning?, © Roslin Institute, Cloning: new horizons in medicine, © Association of the British Pharmaceutical Industry, Human cloning – the ethical issues, © Church of Scotland, A humanist discussion of embryo research, © British Humanist Association, Biologists want to drop the word 'cloning', © Reed Business Information Ltd, Animal-human embryos, © Telegraph Group Ltd, Chimeras, hybrids and 'cybrids', © Christian Medical Foundation, Animal-human hybrids: it makes sense to say no, © CORE, Hybrid embryo ban 'would cost patients' lives', © Telegraph Group Ltd.

Chapter Two: Stem Cell Research

Stem cells and human embryos, © Association of Medical Research Charities, Stem cell research: new horizons in medicine, © Association of the British Pharmaceutical Industry, Frequently asked questions on stem cells, © North East England Stem Cell Institute, Europeans and biotechnology, © European Commission, Genetic engineering, © Guardian Newspapers Ltd, Your stem cell body repair kit, © Associated Newspapers Ltd, Stem cells – too fast too soon?, © Wellcome Trust, An ethical solution to stem cell controversy?, © Massachusetts Institute of Technology, Stem cell milestones, © Medical Research Council, Heart tissue from stem cells, © Technion Israel Institute of Technology, Women may be able to grow own sperm, © Telegraph Group Ltd.

Chapter Three: Animal Cloning

Dolly, © Roslin Institute, Where Dolly went astray, © Guardian Newspapers Ltd, Man or mouse, © Animal Aid, Clones and factory farming, © NewsTarget, Milk from cloned cows leaks into UK, © Associated Newspapers Ltd, Can we be sure no cloned animals are in food chain?, © The Scotsman, Cloned meat and milk 'safe', © Reed Business Information.

Illustrations

Pages 1, 15, 22: Angelo Madrid; pages 6, 20, 30: Simon Kneebone; pages 10, 26: Bev Aisbett; pages 13, 25, 38: Don Hatcher.

Photographs

Page 2: Eyup Salman; page 11: Konstantinas Jariomenko; page 17: Frederic Carmel; page 18: Nick Stuart; page 23: Iwan Beijes; page 26: Joao Estevao; page 37: Samuel Rosa;

And with thanks to the team: Mary Chapman, Sandra Dennis and Jan Haskell.

Lisa Firth and Cobi Smith
Cambridge
September, 2007